MOLECULAR CHEMISTRY AND BIOMOLECULAR ENGINEERING

BIOMOLECULAR ENGINEERING

Integrating Theory and Research with Practice

Innovations in Physical Chemistry: Monograph Series

MOLECULAR CHEMISTRY AND BIOMOLECULAR ENGINEERING

Integrating Theory and Research with Practice

Edited by
Lionello Pogliani, PhD
Francisco Torrens, PhD
A. K. Haghi, PhD

Apple Academic Press Inc.
3333 Mistwell Crescent
Oakville, ON L6L 0A2, Canada

Apple Academic Press Inc.
1265 Goldenrod Circle NE
Palm Bay, Florida 32905, USA

© 2020 by Apple Academic Press, Inc.

First issued in paperback 2021

Exclusive worldwide distribution by CRC Press, a member of Taylor & Francis Group
No claim to original U.S. Government works

ISBN 13: 978-1-77463-468-4 (pbk)
ISBN 13: 978-1-77188-792-2 (hbk)

Library and Archives Canada Cataloguing in Publication

Title: Molecular chemistry and biomolecular engineering : integrating theory and research with practice / edited by Lionello Pogliani, PhD, Francisco Torrens, PhD, A.K. Haghi, PhD.

Names: Pogliani, Lionello, editor. | Torrens, Francisco (Torrens Zaragoza), editor. | Haghi, A. K., editor.

Series: Innovations in physical chemistry.

Description: Series statement: Innovations in physical chemistry: monographic series | Includes bibliographical references and index.

Identifiers: Canadiana (print) 20190141859 | Canadiana (ebook) 20190141883 | ISBN 9781771887922 (hardcover) | ISBN 9780429060649 (ebook)

Subjects: LCSH: Molecular structure. | LCSH: Molecules. | LCSH: Chemistry, Physical and theoretical. | LCSH: Chemical engineering. | LCSH: Molecular biology.

Classification: LCC QD461 .M65 2019 | DDC 541/.22—dc23

Library of Congress Cataloging-in-Publication Data

Names: Pogliani, Lionello, editor. | Torrens, Francisco (Torrens Zaragoza), editor. | Haghi, A. K., editor.

Title: Molecular chemistry and biomolecular engineering : integrating theory and research with practice / edited by Lionello Pogliani, Francisco Torrens, A.K. Haghi.

Other titles: Innovations in physical chemistry.

Description: Oakville, ON ; Palm Bay : Apple Academic Press, [2020] | Series: Innovations in physical chemistry: monograph series | Includes bibliographical references and index. | Summary: "This new volume is devoted to molecular chemistry and its applications to the fields of biology. It looks at the integration of molecular chemistry with biomolecular engineering, with the goal of creating new biological or physical properties to address scientific or societal challenges. It takes a both multidisciplinary and interdisciplinary perspective on the interface between molecular biology, biophysical chemistry, and chemical engineering. Molecular Chemistry and Biomolecular Engineering: Integrating Theory and Research with Practice aims to provide effective support for the development of the laboratory and data analysis skills that researchers will draw on time and again for the practical aspects and also gives a solid grounding in the broader transferable skills. Key features: Provides effective explanations and support for the development of a wide range of laboratory and data analysis skills Gives a solid grounding in the broader transferable skills, which are increasingly necessary Associates key principles and advances with potential applications Provides a contemporary introduction to developing a quantitative understanding of how biological macromolecules behave using classical approaches Focuses on the principles, applications, and limitations of different technologies in molecular chemistry and biomolecular engineering"--Provided by publisher.

Identifiers: LCCN 2019027430 (print) | ISBN 9781771887922 (hardback) | ISBN 9780429060649 (ebook)

Subjects: MESH: Biomedical Research--methods | Biochemical Phenomena | Bioengineering--methods | Plants, Medicinal--chemistry

Classification: LCC R852 (print) | LCC R852 (ebook) | NLM W 20.5 | DDC 610.72/4--dc23

LC record available at https://lccn.loc.gov/2019027430

LC ebook record available at https://lccn.loc.gov/2019027431

Apple Academic Press also publishes its books in a variety of electronic formats. Some content that appears in print may not be available in electronic format. For information about Apple Academic Press products, visit our website at **www.appleacademicpress. com** and the CRC Press website at **www.crcpress.com**

INNOVATIONS IN PHYSICAL CHEMISTRY: MONOGRAPH SERIES

This book series offers a comprehensive collection of books on physical principles and mathematical techniques for majors, non-majors, and chemical engineers. Because there are many exciting new areas of research involving computational chemistry, nanomaterials, smart materials, high-performance materials, and applications of the recently discovered graphene, there can be no doubt that physical chemistry is a vitally important field. Physical chemistry is considered a daunting branch of chemistry—it is grounded in physics and mathematics and draws on quantum mechanics, thermodynamics, and statistical thermodynamics.

Editors-in-Chief

A. K. Haghi, PhD
Editor-in-Chief, *International Journal of Chemoinformatics* and *Chemical Engineering and Polymers Research Journal*; Member, Canadian Research and Development Center of Sciences and Cultures (CRDCSC), Montreal, Quebec, Canada
E-mail: AKHaghi@Yahoo.com

Lionello Pogliani, PhD
University of Valencia-Burjassot, Spain
E-mail: lionello.pogliani@uv.es

Ana Cristina Faria Ribeiro, PhD
Researcher, Department of Chemistry, University of Coimbra, Portugal
E-mail: anacfrib@ci.uc.pt

BOOKS IN THE SERIES

- **Chemical Technology and Informatics in Chemistry with Applications**
  Editors: Alexander V. Vakhrushev, DSc, Omari V. Mukbaniani, DSc, and Heru Susanto, PhD

- **Chemistry and Industrial Techniques for Chemical Engineers**
  Editors: Lionello Pogliani, PhD, Suresh C. Ameta, PhD, and A. K. Haghi, PhD

- **Engineering Technologies for Renewable and Recyclable Materials: Physical-Chemical Properties and Functional Aspects**
  Editors: Jithin Joy, Maciej Jaroszewski, PhD, Praveen K. M., and Sabu Thomas, PhD, and Reza Haghi, PhD

- **Engineering Technology and Industrial Chemistry with Applications**
  Editors: Reza Haghi, PhD, and Francisco Torrens, PhD

- **High-Performance Materials and Engineered Chemistry**
  Editors: Francisco Torrens, PhD, Devrim Balköse, PhD, and Sabu Thomas, PhD

- **Methodologies and Applications for Analytical and Physical Chemistry**
  Editors: A. K. Haghi, PhD, Sabu Thomas, PhD, Sukanchan Palit, and Priyanka Main

- **Modern Green Chemistry and Heterocyclic Compounds: Molecular Design, Synthesis, and Biological Evaluation**
  Editors: Ravindra S. Shinde, and A. K. Haghi, PhD

- **Modern Physical Chemistry: Engineering Models, Materials, and Methods with Applications**
  Editors: Reza Haghi, PhD, Emili Besalú, PhD, Maciej Jaroszewski, PhD, Sabu Thomas, PhD, and Praveen K. M.

- **Molecular Chemistry and Biomolecular Engineering: Integrating Theory and Research with Practice**
  Editors: Lionello Pogliani, PhD, Francisco Torrens, PhD, and A. K. Haghi, PhD

- **Physical Chemistry for Chemists and Chemical Engineers: Multidisciplinary Research Perspectives**
  Editors: Alexander V. Vakhrushev, DSc, Reza Haghi, PhD, and J. V. de Julián-Ortiz, PhD

- **Physical Chemistry for Engineering and Applied Sciences: Theoretical and Methodological Implication**
  Editors: A. K. Haghi, PhD, Cristóbal Noé Aguilar, PhD, Sabu Thomas, PhD, and Praveen K. M.

- **Research Methodologies and Practical Applications of Chemistry**
  Editors: Lionello Pogliani, PhD, A. K. Haghi, PhD, and Nazmul Islam, PhD

- **Theoretical and Empirical Analysis in Physical Chemistry: A Framework for Research**
  Editors: Miguel A. Esteso, PhD, Ana Cristina Faria Ribeiro, PhD, and A. K. Haghi, PhD

- **Theoretical Models and Experimental Approaches in Physical Chemistry: Research Methodology and Practical Methods**
  Editors: A. K. Haghi, PhD, Sabu Thomas, PhD, Praveen K. M., and Avinash R. Pai

CONTENTS

ABOUT THE EDITORS

Lionello Pogliani, PhD
University of Valencia-Burjassot, Spain
E-mail: lionello.pogliani@uv.es

Lionello Pogliani, PhD, is a retired professor of physical chemistry. He received his postdoctoral training at the Department of Molecular Biology of the C. E. A. (Centre d'Etudes Atomiques) of Saclay, France, at the Physical Chemistry Institute of the Technical and Free University of Berlin, and at the Pharmaceutical Department of the University of California, San Francisco, USA. He spent his sabbatical years at the Technical University of Lisbon, Portugal, and at the University of Valencia, Spain. He has contributed more than 200 papers in the experimental, theoretical, and didactical fields of physical chemistry, including chapters in specialized books and a book on numbers 0, 1, 2, and 3. A work of his has been awarded with the GM Neural Trauma Research Award. He is a member of the International Academy of Mathematical Chemistry and he is on the editorial board of many international journals. He is presently a part-time teammate at the Physical Chemistry Department of the University of Valencia, Spain.

Francisco Torrens, PhD
Lecturer, Physical Chemistry, Universitat de València, València, Spain

Francisco Torrens, PhD, is lecturer in physical chemistry at the Universitat de València in Spain. His scientific accomplishments include the first implementation at a Spanish university of a program for the elucidation of crystallographic structures and the construction of the first computational-chemistry program adapted to a vector-facility supercomputer. He has written many articles published in professional journals and has acted as a reviewer as well. He has handled 26 research projects, published two books and over 350 articles, and made numerous presentations.

A. K. Haghi, PhD

Professor Emeritus of Engineering Sciences, Former Editor-in-Chief, International Journal of Chemoinformatics and Chemical Engineering and Polymers Research Journal; Member, Canadian Research and Development Center of Sciences and Cultures (CRDCSC), Canada

A. K. Haghi, PhD, is the author and editor of 165 books, as well as 1000 published papers in various journals and conference proceedings. Dr. Haghi has received several grants, consulted for a number of major corporations, and is a frequent speaker to national and international audiences. Since 1983, he served as professor at several universities. He was formerly the editor-in-chief of the *International Journal of Chemoinformatics and Chemical Engineering* and *Polymers Research Journal* and on the editorial boards of many international journals. He is also a member of the Canadian Research and Development Center of Sciences and Cultures (CRDCSC), Montreal, Quebec, Canada.

CONTRIBUTORS

Ricardo Aguilar-López
Departamento de Biotecnología y Bioingeniería, CINVESTAV-IPN,
Av. Instituto Politécnico Nacional No. 2508, Col. San Pedro Zacatenco, 07360,
Mexico City, CDMX., Mexico

Gloria Castellano
Departamento de Ciencias Experimentales y Matemáticas, Facultad de Veterinaria y Ciencias
Experimentales, Universidad Católica de Valencia San Vicente Mártir, Guillem de Castro-94,
E-46001 València, Spain

Kaushal Chaturvedi
Lachman Institute for Pharmaceutical Analysis Laboratory, Long Island University, New York, USA

Pankaj Dangre
R.C. Patel Institute of Pharmaceutical Education and Research, Shirpur, India.
E-mail: pankaj_dangre@rediffmail.com

Pablo A. López-Pérez
Escuela Superior de Apan, Universidad Autónoma de Hidalgo, Carretera Apan-Calpulalpan,
km. 8. Chimalpa Tlalayote s/n Colonia Chimalpa, Apan, 43900 Hgo. México.
E-mail: vicente.caballero@ugto.mx

V. Peña-Caballero
Departamento de Ingeniería Agroindustrial, División de Ciencias de la Salud e Ingenierías.
Campus Celaya-Salvatierra, Universidad de Guanajuato, Ave. Ing. Barros Sierra No. 201,
Esq. Ave. Baja California. Ejido de Santa María del Refugio, C.P. 38140, Mexico.
E-mail: sasve1991@yahoo.com.mx

Dimitar N. Petrov
Department of Physical Chemistry, Faculty of Chemistry, Paisii Hilendarski University,
24 Tzar Asen Str., 4000 Plovdiv, Bulgaria. E-mail: petrov_d_n@abv.bg

Prasad Pofali
Shobhaben Pratapbhai Patel School of Pharmacy and Technology Management,
SVKM's NIMMS, Mumbai, India

Lionello Pogliani
Department de Química Física, Facultad de Farmacia, Universitat de Valencia,
Av. V.A. Estellés s/n, 46100 Burjassot, València, Spain

Francisco Torrens
Institut Universitari de Ciència Molecular, Universitat de València, Edifici d'Instituts de Paterna,
P. O. Box 22085, E-46071 València, Spain. E-mail: torrens@uv.es

ABBREVIATIONS

ACh	acetylcholine
AEC	Atomic Energy Commission
AFM	atomic force microscopy
AFP	*Atoms for Peace*
APIs	active pharmaceutical ingredients
APP	acute postoperative pain
ART	artemisinin
As	answers
BC	breast cancer
BCS	Biopharmaceutical Classification System
BSO	L-buthionine sulphoximine
BTX	botulinum toxin
CBT	cognitive–behavioral therapy
CF	crystal field
CM	conventional medical
CN	coordination number
COX-2	cyclooxygenase-2
CRT	cathode ray tube
CSE	change of a standard enthalpy
CSTR	continuous flow stirred-tank reactor
CT	charge transfer
DFT	density functional theory
DL	drug loading
DNA	deoxyribonucleic acid
DPPH	2,2-diphenyl-1-picrylhydrazyl
DSC	differential scanning calorimetry
DSSC	dye-sensitized solar cells
EOs	essential oils
EPBs	extracellular polymeric substances
FAP	familial adenomatous polyposis
FITC	fluorescein isothiocyanate
FLU	fluorescence
FRET	Förster resonance energy transfer

FWHM	full width at half maximum
GC-FID	gas chromatography/flame ionization detector
GRAS	generally regarded as safe
GSH	glutathione
HeLa	Henrietta Lacks
HF	health folder
Hs	hypotheses
IAEA	International Atomic Energy Agency
IL	interleukin
iNOS	inducible NO synthase
ITS	individual temporary store
iv	intravenous
LA	living agreement
LD	laser diffraction
LiLnTP	lithium lanthanide tetraphosphates
Ln	lanthanide element
$LnAlO_3$	lanthanide monoaluminates
LNs	lipidic nanoparticles
LPS	lipopolysaccharide
LSD	lysergsäure-diethylamid
ML	α-methylene-γ-lactone
MP	Manhattan Project
MS	mass spectrometry
NAC	N-acetylcysteine
nAChR	nicotinic acetylcholine receptor
NCLs	nanostructured lipid carriers
NF-κB	nuclear factor-κB
NGO	nongovernmental organization
NGS	next-generation sequencing
NO	nitric oxide
NPs	nanoparticles
NSCs	neural stem cells
OLA	orthorhombic lanthanide monoaluminates
OPA	orthophthaldehyde
PC	point charge
PCS	photon correlation spectroscopy
PDI	polydispersity index
PG	prostaglandin

PP	postoperative pain
PS	phosphatidylserine
PV	photovoltaic
PXRD	powder X-ray diffraction
QLS	quasi-elastic light scattering
Qs	questions
RS	Russell–Saunders
RT	room temperature
SEM	scanning electron microscopy
SLNs	solid-lipid nanoparticles
SO	spin–orbit
SRB	sulfate-reducing bacteria
STLs	sesquiterpene lactones
STO	Slater type orbital
TCM	traditional Chinese medicine
TCOI	two-center overlap integral
TEM	transmission electron microscopy
TNF-α	tumor necrosis factor-α
TSRS	two-stage reactor system
UN	United Nations
UV–VIS	UV–visible
VUV	vacuum ultraviolet
WW2	World War II
YAG	yttrium aluminum garnet

PREFACE

Modern science is continuously leaning toward chemistry. This book is devoted to molecular chemistry and its applications to the fields of biology. Molecular chemistry is a creative science where chemists synthesize molecules with new biological or physical properties to address scientific or societal challenges.

Biomolecular engineering is an emerging discipline at the interface of molecular biology, biophysical chemistry, and chemical engineering—whose express purpose is developing novel molecular tools and approaches that are the focal point of applied and basic research within academia, industry, and medicine. This effort includes both multidisciplinary and interdisciplinary activities, including applied mathematics and physical chemistry.

This book focuses on the interface between molecular biology, biophysical chemistry, and chemical engineering. In the first chapter, two nonconventional chemical kinetics approaches for writing and solving the rate laws are introduced and discussed. The first approach is based on matrix algebra, and the second one on a direct integrated form, known as convolution method.

In Chapter 2, analysis of chemical and biological kinetic reactor systems to remove hexavalent chromium via sulfate-reducing process is investigated in detail.

Fluorescence exitation and emission: comparison with absorption is discussed in Chapter 3.

The objective of Chapter 4 is to review ethnobotanical studies of medicinal plants, underutilized wild edible plants, food, and medicine; bioactive compounds in *Zingiber zerumbet* rhizomes essential oils (EOs); antiallergic and immunomodulatory properties, and the ethnopharmacology of love.

The chemical components from *Artemisia austro-yunnanensis* anti-inflammatory effects and lactones are presented in Chapter 5.

Chapter 6 deals with stoichiometric lanthanide and doped rare-earth complex oxides. It contains a survey of recent studies on certain basic properties and applications of 4fn electron systems. The following topics

are included: lattice energies, dielectric properties, magnetic susceptibilities and magnetic exchange interactions, optical spectra, nephelauxetic
effect and derivation of radial expectation values for Ln^{3+} ions, two-center
overlap with participation of 4f electron wave functions, and crystalline
materials of nanosized particles.

The main aim of Chapter 7 is to review the biological activities and
safety of *Citrus* spp. EOs.

Health in solitude and thrill of fragility is discussed in Chapter 8.

The aim of Chapter 9 is to initiate a debate by suggesting a number of
questions that can arise when addressing subjects of cancer and hypotheses on cancer in different fields, and providing, when possible, answers
and hypotheses.

Chapter 10 initiates a debate by suggesting a number of questions that
can arise when addressing subjects of pain and pleasure in different fields,
and providing, when possible, answers, hypotheses, and conclusions. Pain,
*an unpleasant sensory and emotional experience associated with actual
or potential tissue damage, or described in terms of such damage*, acts
undoubtedly as an alarm system, and the integrity of the organism depends
on its correct functioning.

In Chapter 11, new insights on lipidic nanoparticles as a platform for
advancement in drug delivery system are presented in detail.

The aim of Chapter 12 is to initiate a debate by suggesting a number of
questions (Q) that can arise when addressing subjects of *Atoms for Peace*
in different fields, and providing, when possible, answers (A), hypotheses
(H), and paradoxes (P).

CHAPTER 1

SOME REMARKS ABOUT MATRIX AND CONVOLUTION KINETICS

LIONELLO POGLIANI*

Department de Química Física, Facultad de Farmacia, Universitat de Valencia, Av. V.A. Estellés s/n, 46100 Burjassot, València, Spain

E-mail: liopo@uv.es

ABSTRACT

Two nonconventional chemical kinetics approaches for writing and solving the rate laws are introduced and discussed. The first approach is based on matrix algebra, and the second one on a direct integrated form, known as convolution method. Matrix algebra offers an elegant way for solving the differential equations for simple kinetic systems as it allows the simultaneous solution for the reaction rates for the different chemical species. Thanks to a set of rules, this method allows deriving the kinetic rate matrix of kinetic systems of any kind in a direct and straightforward way, and it is particularly advantageous for first-order steps. The convolution approach applies to species that are consumed solely through first-order steps, regardless of the complexity of its formation pathways. This second method offers the possibility to formulate the rate equation directly in the integrated form, a form that shows an interesting structure, especially in the case of consecutive reaction schemes.

1.1 INTRODUCTION

The evaluation of the kinetic rate constants and order for a given reaction scheme is usually done by a fitting procedure of the experimental results to the integrated kinetic equation. For this reason, an important part of lectures on chemical kinetic is focused on the integration of the corresponding rate laws that are expressed as differential equations.[1-3] Normally, this topic consists in studying the fate of a specific reagent or product along several specific cases, that is, first-order, first-order opposing or reversible and consecutive, second-order, second-order reversible, etc. This "state of the art" gives the impression of a lack of unity and leaves the feeling that each case is somewhat unique.

The use of matrices in chemical kinetics allows the formulation of chemical models in an elegant and compact way. Nevertheless, the integration of kinetic rate equations is usually presented without recourse to matrix algebra. The matrix formulation of the rate equations (a set of interconnected differential equations, each displaying the concentration time variation of a specific species) is particularly convenient since it allows an indirect integration of the rate equations, using a uniform set of procedures. In addition, the time evolution of the concentrations of all species (reagents, products, and intermediates) is obtained simultaneously. This topic that is only briefly treated in basic textbooks on chemical kinetics and mathematics for chemistry [4-6] has been reconsidered more than 20 years ago and further developed by different authors.[7-12] The application of matrix algebra to kinetics is not restricted to first-order elementary steps, where the definition of a set of rules that can help to formulate the rate K matrix is rather straightforward, but also to higher-order kinetics. Actually, numerical methods based on the same algebraic approach can be used to solve the more general case of kinetic systems composed by steps of any order. The reader should notice that the algebraic formalism centered on the use of matrices is helpful in other well-known branches of chemistry and chemical engineering, like quantum chemistry (secular equation), spectroscopy (molecular vibrations), and in a less known branch known as chemical graph theory,[13] and its applications to chemical kinetics.[14,15]

The second part of the paper is devoted to the convolution approach that allows writing kinetic equations directly in the integrated form. This approach is limited to kinetic schemes composed of first-order or pseudo first-order elementary steps[12,16-22] and it has been used for the analysis of complex photochemical kinetic systems.

1.2 RESONANCE OF THE TWO METHODS

Let us shortly turn our attention to the reception that these two methods received throughout the chemical literature. We will here cite some of the studies that got interested in the two kinetic methods. Let us start considering what seems to be the more cited subject, the matrix kinetics. Many chemical studies[23–35] have used the mathematical technique of matrix algebra that was explained in the two *Journal of Chemical Education* papers of 1990 and 1992[7,8] that were both well-received thanks to their didactic character. The reader can notice the echo of these two studies stretched over a wide range of chemical and biological fields: chemical kinetics, educational chemistry, physical chemistry, molecular biology, biophysics, biochemistry, and material chemistry. Worth citing is also a book that teaches how to use matrix algebra with Mathcad and Maple[36] softwares.

Convolution kinetics has caught the attention of a minor number of specialists as can be seen from the cited literature,[34,35,37–50] even if it is frequent the case of studies that expound both kinetic approaches as taught in Refs. [19] and [20]. Actually, these two papers seem to have been quite successful in "advertising" the two methods, and for this reason, throughout the following section they will be used as a guideline. From the cited References, it is also possible to notice that the convolution method in chemical kinetics caught the attention of investigators mainly interested in the mathematical characteristics of the method even if it also raised the attention of physical, biological chemists, and engineers.

1.3 OUTLINE OF THE TWO MATHEMATICAL METHODS

1.3.1 THE MATRIX KINETICS

Be three species A, B, and C that take part to a first- or pseudo first-order chemical kinetic process. The more general first-order K rate matrix that describes this kinetic system is the following 3×3 square matrix (actually the transpose of a normal mathematical matrix),

$$\begin{bmatrix} -k_{AA} & k_{BA} & k_{CA} \\ k_{AB} & -k_{BB} & k_{CB} \\ k_{AC} & k_{BC} & -k_{CC} \end{bmatrix} = K \tag{1.1}$$

The minus sign along the main diagonal means that along that path, species undergo chemical consumption. The k_{AA}, k_{BB}, k_{CC} terms are the rate constants for reactions departing from reactants A, B, and C, that is, $\leftarrow A \rightarrow$, $\leftarrow B \rightarrow$, and $\leftarrow C \rightarrow$. For instance, k_{AA} is the sum of the rate constants for $A \rightarrow B$ and $A \rightarrow C$ reactions. A cross-term, k_{IJ}, is the rate constants for the only process $I \rightarrow J$. Thus, k_{AB} and k_{BA} are the rate constants for the reactions $A \rightarrow B$ and $B \rightarrow A$.

These K matrices share two properties:[9,11] (i) the sum of terms along a column is always zero as the term on the main diagonal is the negative sum of the terms along its column, (ii) for reversible reactions, the terms on symmetric sides of the main diagonal differ each other by their direction (either forward or reverse). In matrix kinetics, the rate equation can be written in the following succinct way, where $d[X]/dt = dX/dt = X'$, and X stands for a nonspecific concentration,

$$X' = K \times X \tag{1.2}$$

Here, X' is the vector of the kinetic rates, and X is the vector of the concentrations. Let us now apply the two rules to build the K matrices for two first-order reaction schemes.

1.3.1.1   CONSECUTIVE FIRST-ORDER REACTIONS

$A \rightarrow B \rightarrow C$

Here, we have $k_{AA} = k_{AB} = k_1$ and $k_{BB} = k_{BC} = k_2$ (there is no reverse reaction from B to A and C to B), while all other k's $= 0$. The K matrix here is,

$$\begin{bmatrix} -k_1 & 0 & 0 \\ k_1 & -k_2 & 0 \\ 0 & k_2 & 0 \end{bmatrix} = K \tag{1.3}$$

Notice that the property (i) stands, while the property (ii) is meaningless as the reaction is not reversible.

1.3.1.2 OPPOSING FIRST-ORDER CONSECUTIVE REACTIONS

$$A \underset{k_1}{\overset{k_{-1}}{\rightleftarrows}} B \underset{k_2}{\overset{k_{-2}}{\rightleftarrows}} C \underset{k_3}{\overset{k_{-3}}{\rightleftarrows}} D$$

Here, $k_{AA} = k_{AB} = k_1$, $k_{BA} = k_{-1}$, $k_{BB} = (k_{-1} + k_2)$, $k_{BC} = k_2$, $k_{CB} = k_{-2}$, $k_{CC} = (k_{-2} + k_3)$, $k_{CD} = k_3$, $k_{DC} = k_{DD} = k_{-3}$, all other k's = 0. The rate matrix is the following: a 4×4 matrix, where it is easy to notice that properties (i) and (ii) hold,

$$\begin{bmatrix} -k_1 & k_{-1} & 0 & 0 \\ k_1 & -(k_{-1} + k_2) & k_{-2} & 0 \\ 0 & k_2 & -(k_{-2} + k_3) & k_{-3} \\ 0 & 0 & k_3 & -k_{-3} \end{bmatrix} = K \qquad (1.4)$$

1.3.1.3 SOLUTION OF A FIRST-ORDER CONSECUTIVE REACTION SCHEME

Let us consider the case of the consecutive first-order reaction: $A \to B \to C$. This system by the aid of the matrix eigenvalue method is amenable to a closed form solution.[4,7,8] The $X' = K \times X$ matrix equation, with: $X' = (A', B', C')$ and $X = (A, B, C)$, has the following general solution,

$$X = \begin{bmatrix} K_1 \\ K_2 \\ K_3 \end{bmatrix} \exp(\lambda t) \qquad (1.5)$$

With this solution, eq 1.2 led us to the following matrix eigenvalue problem, as $X' = \lambda X$:

$$\lambda X = K \times X \qquad (1.6)$$

as $\lambda X = \lambda IX$, where I is the unit matrix, rearranging we obtain,

$$(K - \lambda I) \times X = 0 \tag{1.7}$$

For nontrivial solutions to this equation to exist, the determinant $\Delta = |K - \lambda I|$ must be zero. Solution for $\Delta = 0$ yields the eigenvalues: $\lambda_1 = 0$, $\lambda_2 = -k_1$, and $\lambda_3 = -k_2$. The X_1, X_2, and X_3 eigenvectors belonging to these eigenvalues are obtained with the matrix-column vector product of eq 1.8. They are displayed in eq 1.9, where $\Delta k = k_2 - k_1$ and L, M, and N are three non-zero constants.

$$\begin{bmatrix} -(k_1 - l) & 0 & 0 \\ k_1 & -(k_2 - l) & 0 \\ 0 & k_2 & -l \end{bmatrix} \begin{bmatrix} K_1 \\ K_2 \\ K_3 \end{bmatrix} = 0 \tag{1.8}$$

$$X_1 \begin{bmatrix} 0 \\ 0 \\ L \end{bmatrix} X_2 \begin{bmatrix} M \\ Mk_1 / Dk \\ -Mk_2 / Dk \end{bmatrix} \exp(-k_1 t) - X_3 = \begin{bmatrix} 0 \\ N \\ -N \end{bmatrix} \exp(-k_2 t) \tag{1.9}$$

The general solution of the consecutive reaction scheme involving only irreversible first-order steps is the linear combination of the three independent X_1, X_2, and X_3 eigenvectors as shown in eq 1.10. The initial conditions for $t = 0$: $A(0) = A_0$, $B(0) = B_0$, and $C(0) = C_0$ allow to get the values for L, M, and N.

$$X = \begin{bmatrix} A \\ B \\ C \end{bmatrix} = \begin{bmatrix} Mexp(-k_1 t) \\ Mk_1 exp(-k_1 t) / Dk & + & Nexp(-k_2 t) \\ L & & - Mk_2 exp(-k_1 t) / Dk & - Nexp(-k_2 t) \end{bmatrix} \tag{1.10}$$

The final time-dependent concentrations for the three species with $S = A_0 + B_0 + C_0$ are:

$$A = A_0 exp(-k_1 t) \tag{1.10a}$$

$$B = \frac{1}{\Delta k} \{ A_0 k_1 \exp \exp(-k_1 t) + [B_0 \Delta k - A_0 k_1] \exp(-k_2 t) \} \tag{1.10b}$$

$$C = S - \frac{1}{\Delta k} \{ A_0 k_2 \exp \exp(-k_1 t) - [B_0 \Delta k - A_0 k_1] \exp(-k_2 t) \} \tag{1.10c}$$

For $t \to \infty$, we have, as expected, $A = B = 0$ and $C = S$. The reader should check for $t = 0$.

1.3.1.4 THE SECOND-ORDER KINETIC MATRIX

The more general case of a 4×4 matrix for second-order kinetics with four species is shown in eq 1.11. Modifications relatively to the first-order rate matrices are due to the introduction of a concentration term $(X)^{n-1}$ that multiplies the k_{IJ} rate constants. Here, $n = 2$, $X = (A, B, C, D)$, and the concentration terms are here inverted, that is, $X_r = (B, A, D, C)$.

$$\begin{bmatrix} -k_{AA}B & k_{BA}A & k_{CA}D & k_{DA}C \\ k_{AB}B & -k_{BB}A & k_{CB}D & k_{DB}C \\ k_{AC}B & k_{BC}A & -k_{CC}D & k_{DC}C \\ k_{AD}B & k_{BD}A & k_{CD}D & -k_{DD}C \end{bmatrix} = K \qquad (1.11)$$

This matrix can be converted into a pseudo-first-order matrix for $n = 1$, as $(X)^{n-1} = (X)^0 = (B^0, A^0, D^0, C^0) = (1, 1, 1, 1)$.

To be more specific, and more concrete, let us consider the following bimolecular second-order reaction: $A + B \rightleftharpoons C + D$. Equation 1.11 can be simplified in two steps:

(1) As there is no internal reaction between reactants or products (i.e., A and C do not result from B and D, respectively, and vice versa), we have $k_{AB} = k_{BA} = k_{CD} = k_{DC} = 0$.

(2) As third and fourth term in the first and second rows as well as first and second term in the third and fourth row is redundant, one of them should be deleted. The redundancies can be detected looking for terms in a row[9,11] (excluding terms in the main diagonal) that differ each other by internal exchange of reactants or products (namely, by internal exchange of subscripts). The resulting matrix will then be:

$$\begin{bmatrix} -k_{AA}B & 0 & 0 & k_{DA}C \\ 0 & -k_{BB}A & k_{CB}D & 0 \\ 0 & k_{BC}A & -k_{CC}D & 0 \\ k_{AD}B & 0 & 0 & -k_{DD}C \end{bmatrix} = K \qquad (1.12)$$

Redundancies were eliminated in order to obtain both a nice-looking matrix that obeys steps 1 and 2. With the given method to construct second- and first-order matrices, it is possible to build rate matrices of complex kinetic mechanisms composed solely by second- and first-order reaction steps.

1.3.1.5   THE MIXED-ORDER KINETIC MATRIX

Be the following mixed reaction scheme made of second- and first-order elementary steps,

$$A \underset{k_1}{\overset{k_{-1}}{\longrightarrow}} B \underset{k_2}{\overset{k_{-2}}{\longrightarrow}} C$$

$$A \overset{k_3}{\longrightarrow} D$$

$$D + B \overset{k_4}{\longrightarrow} E$$

$$E \underset{k_5}{\overset{k_{-5}}{\longrightarrow}} A + D$$

After elimination of redundancies (last row: first $k_{-5}D$ term and fourth k_4B term),[7,9,12] we obtain the rate matrix of eq 1.13, with concentration vector $X = (A, B, C, D, E)$. Mixed order matrices do not obey anymore rules (i) and (ii). The solution of this kinetic problem is rather formidable and either approximations (such as the steady state or the pre-equilibrium approximations) or numerical methods (such as Euler's or the Runge-Kutta's method) have to be invoked to solve it.[7,20] The concentration vector here is $X = (A, B, C, D, E)$. Mixed order matrices do not obey anymore rules (i) and (ii).

$$
\begin{bmatrix}
-(k_1+k_3+k_{-5}D) & k_{-1} & 0 & 0 & k_5 \\
k_1 & -(k_{-1}+k_2+k_4D) & k_{-2} & 0 & 0 \\
0 & k_2 & -k_{-2} & 0 & 0 \\
k_3 & 0 & 0 & -(k_4B+k_{-5}A) & k_5 \\
0 & k_4D & 0 & k_{-5}A & -k_5
\end{bmatrix} = K \quad (1.13)
$$

The solution of this kinetic problem is rather formidable and either approximations (such as the steady-state or the pre-equilibrium approximations) or numerical methods (such as Euler's or the Runge-Kutta's method) have to be invoked to solve this kinetic problem[7].

1.3.1.6   OTHER TYPES OF RATE MATRICES

The second-order matrix of eq 1.11 can be used as starting point for the construction of other type of matrices such as matrices of second-order steps with stoichiometric coefficients $v_i \neq 1$ or matrices of autocatalytic steps and, clearly, the corresponding matrices of mixed reaction steps. Let us substitute in matrix 1.12 D with C (also throughout the subscripts) and obtain:

$$
\begin{bmatrix}
-k_{AA}B & 0 & 0 & k_{CA}C \\
0 & -k_{BB}A & k_{CB}C & 0 \\
0 & k_{BC}A & -k_{CC}C & 0 \\
k_{AC}B & 0 & 0 & -k_{CC}C
\end{bmatrix} = K \quad (1.14)
$$

This matrix can be transformed into the following one:

$$
\begin{bmatrix}
-k_{AA}B & 0 & k_{CA}C \\
0 & -k_{BB}A & k_{CB}C \\
k_{AC}B & k_{BC}A & -2k_{CC}C
\end{bmatrix} = K \quad (1.15)
$$

If we (1) add fourth column into the third one and delete it, (2) add now the fourth row into the third one and delete it, and, finally, (3) delete the redundant term in the third row (there is a redundancy between first and second term of this row: internal reactant exchange). Thus, we obtain the rate matrix of the following elementary step with stoichiometric coefficient, $v_C = 2$

$$A + B \xrightarrow{\longleftarrow} 2\,C$$

$$\begin{bmatrix} -k_{AA}B & 0 & k_{CA}C \\ 0 & -k_{BB}A & k_{CB}C \\ k_{AC}B & 0 & {}^-2k_{CC}C \end{bmatrix} \tag{1.16}$$

Thus, a way to construct rate matrices of reaction steps with nonunitary stoichiometric is: (1) to expand by the aid of dummy species, the dimension space of the reaction into a space where no equal species are present and (2) to reduce it back into the normal dimension of the reaction operating on the columns and rows of the dummy species.

This process was introduced to handle matrices of autocatalytic reaction steps[9,11] strongly reminds the simplex method for solving linear programing problems. Let us then build a rate matrix for the following autocatalytic step (k_1: forward direction, k_{-1}: reverse direction),

$$A + B \xrightarrow[k_1]{\longleftarrow k^{-1}} 2\,A$$

Let us start with matrix 1.12, where the k_{ij} of the dummy reaction scheme

$$A + B \xrightarrow{\longleftarrow} C + D$$

are substituted with the corresponding k_1 and k_{-1} to avoid errors handling the matrix

$$\begin{bmatrix} -k_1B & 0 & 0 & k_{-1}C \\ 0 & -k_1A & k_{-1}D & 0 \\ 0 & k_1A & -k_{-1}D & 0 \\ k_1B & 0 & 0 & -k_{-1}C \end{bmatrix} = K \tag{1.17}$$

Now, (1) replacing C, and D by A, (2b) adding third and fourth column into the first column and eliminating them, and (3) adding the resulting third and fourth rows into the first row and eliminating them, we obtain the rate matrix (with no redundant terms) of the autocatalytic step that obeys rules (i) and (ii) with $X = (A, B)$.

$$\begin{bmatrix} -k_{-1}A & k_1A \\ k_{-1}A & -k_1A \end{bmatrix} = K \qquad (1.18)$$

The considered processes that allow starting with a general second-order K matrix and derive specific K matrices of the same or lower order can be applied to any n^{th} order K matrix from which it should be possible to derive any matrices of lower order. The same matrix formalism, with minor changes, can also be applied to derive kinetic K matrices in open systems (in continuous flow stirred-tank reactor, i.e., CSTR) .[9,11] In this case, two new terms have to be added in each term of the main diagonal: $\Phi\Delta$ and $D_I\nabla^2$, where Φ is the flow rate, Δ is the subtracting operator, that is, $\Delta A = A_0 - A$ (A_0 being the concentration of the input flow), D_I is the diffusion coefficient of species, I and ∇^2 is the laplacian operator. This case will be treated more in detail in a next paragraph.

1.3.1.7   EULER'S NUMERICAL METHOD

This method is also valid for time-dependent $K(t)$ [with $k = k(t)$], which solves the matrix equation

$$X' = K(t) \times X \qquad (1.19)$$

Now, approximating $X' = dX/dt$ with $\Delta X/\Delta t$

$$\Delta X / \Delta t = [X(t + \Delta t) - X(t)] / \Delta t = K(t) \times X(t) \qquad (1.19)$$

Rearranging $[X(t + \Delta t) = X(t) + K(t) \times X(t) \times \Delta t]$ we obtain eq 1.19, where U is the unit matrix of order n,

$$X(t + \Delta t) = [U + K(t) \Delta t] X(t) \qquad (1.19)$$

Repeated applications of eq 1.19 (with constant or variable time increment Δt and assuming that $X(0)$ at $t = 0$ is known) allows the calculation of the concentration of any species at any instant.

1.3.2   CONVOLUTION KINETICS

Macroscopic chemical kinetics is based on differential equations of the type

$$dC_i / dt = \Sigma_{k = i, o, p, c} (dC_i / dt)_k \qquad (1.20)$$

They are simple balances for the amount of species C_i ($i = A, B, ...$) within the system. The sum runs over the four main processes occurring in a chemical reaction: i stands for input, o for output, p for internal production, and c for internal consumption. The first two processes being relevant for open systems and the last two being associated with chemical reactions occurring within the system. Comparison with experimental data is usually done in the integrated form, that is, integrating eq 1.20 analytically or numerically. The resulting function C_i (t) is then checked with the experimental data in order to extract rate constants and confirm the proposed mechanism. Integration of eq 1.20 can be made with several mathematical techniques, including the discussed matrix methods.

Actually, when all the consumption rates are first order, it is possible to write down the balances directly in the integrated form. Suppose that a reactive chemical species C_i can be instantaneously produced at unit concentration at time $t = 0$; ignoring the possibility of reformation of C_i via closed loops (e.g., a reversible step), its time evolution will be given by a certain function $C_i(t)$. This function is the response to a unit input of C_i at time zero, that is, to a Dirac's delta function (t), and reflects all possible disappearance routes for C_i. Under first-order or pseudo-first order conditions, this function is independent of all concentrations and is given by eq 1.21, where k_{ij} are strict first or pseudo-first-order rate constants of the elementary steps by which C_i disappears.

$$C_{i\delta}(t) = \exp(-\Sigma_j k_{ij} t) \qquad (1.21)$$

The general time evolution for C_i is then given by the following convolution integral,

$$C_i(t) = \int_0^t P_i(\theta) C_{i\delta}(t - \theta) d\theta = P_i(t) \otimes C_{i\delta}(t) \qquad (1.22)$$

To solve the convolution integral, use of the Laplace transform is required.[51,52] This transform is a linear operator L, (i.e., $L[\Sigma a_i f_i] = \Sigma a_i L[f_i]$), which transforms a differential equation into an algebraic one, changing the space in which the function is defined, and has the interesting property to transform a convolution product into the product of the individual transforms (i.e., $L[f \otimes g] = L[f] \times L[g]$). Both properties (linearity and convolution handling) are shared by the respective anti-Laplacian L^{-1} operator, which transforms $F(s)$ back to $f(t)$, that is, $f(t) = L^{-1}[F(s)]$. A rather easy convolution problem is represented by the case, $P(t) = \exp(-at)$ and $C_{i\delta}(t) = \exp(-bt)$, as in this case the convolution integral can directly be solved (see following consecutive reaction section), the final solution being $[\exp(-at) - \exp(-bt)]/(b - a)$. Furthermore, an isolated chemical species A can be imagined as the output of an unitary Dirac-like pulse with $P(t) = N \otimes \delta(t)$, and $C_{A\delta}(t) = 1$, and given by the equation: $C_A(t) = P(t) \otimes C_{A\delta}(t) = N \otimes \delta(t) \otimes \times 1 = N$, where the normalization constant N is the concentration of A at $t = 0$, that is, $N = A_0$.

Summing up, for a given kinetic problem, the full solution in terms of the convolution approach is obtained in four steps: (1) identification of the delta responses $C_{i\delta}(t)$, (2) identification of the production terms $P_i(t)$; (3) writing of the system of coupled (through the P_i's) integral equations $C_i(t) = P_i(t) \otimes C_{i\delta}(t)$ ($i = A, B, \ldots$), and (4) solution of the system of coupled equations, for example, with Laplace transform. Some cases will now be discussed.

1.3.2.1 CONSECUTIVE FIRST-ORDER REACTIONS

$$A \overset{k_1}{\to} B \overset{k_2}{\to} C$$

Let the initial concentrations of A, B, and C be A_0, 0, and 0, respectively. The time evolution of A, B, and C in response to a δ-production of each species are dictated by their routes of disappearance and production rates (P),

$$A_\delta(t) = \exp(-k_1 t) \tag{1.23}$$

$$B_\delta(t) = \exp(-k_2 t) \tag{1.24}$$

$$C_\delta(t) = 1 \tag{1.25}$$

$$P_A = A_0 \delta(t) \tag{1.26}$$

$$P_B = k_1 A \tag{1.27}$$

$$P_C = k_2 B \tag{1.28}$$

The system of coupled integral equations, $C_i(t) = P_i(t) \otimes C_{i\delta}(t)$, here is,

$$A = P_A(t) \otimes A_\delta(t) = A_0 \delta(t) \otimes \exp(-k_1 t) = A_0 \exp(-k_1 t) \tag{1.29}$$

$$B = k_1 A \otimes B_\delta(t) = k_1 A_0 \exp(-k_1 t) \otimes \exp(-k_2 t) \tag{1.30}$$

$$C = k_2 B \otimes C_\delta(t) = k_2 k_1 A_0 \exp(-k_1 t) \otimes \exp(-k_2 t) \otimes 1 \tag{1.31}$$

From the definition of convolution and performing the convolution integrals (as already told, here the convolution integral can directly be solved), one obtains, with $\Delta k = k_2 - k_1$,

$$\exp(-k_1 t) \otimes \exp(-k_2 t) = \{\exp(-k_1 t) - \exp(-k_2 t)\} / \Delta k \tag{1.32}$$

Now, with $k_1 = 0$, and with the commutativity property of convolution, we obtain

$$\exp(-k_2 t) \otimes 1 = \{1 - \exp(-k_2 t)\} / k_2 \tag{1.33}$$

After substitution, we have:

$$\exp(-k_1 t) \otimes \exp(-k_2 t) \otimes 1 = \{\exp(-k_1 t) \otimes 1 - \exp(-k_1 t) \otimes \exp(-k_2 t)\} / k_2 \tag{1.34}$$

Finally, we obtain the result for the time evolution for the three species:

$$A = A_0 \exp(-k_1 t) \tag{1.35a}$$

$$B = \frac{A_0 k_1}{\Delta k} \left[\exp(-k_1 t) - \exp(-k_2 t) \right] \tag{1.35b}$$

$$C = A_0 \left[1 + \frac{k_1 \exp \exp(-k_2 t) - k_2 \exp(-k_1 t)}{\Delta k} \right] \tag{1.35c}$$

Note that for non-zero initial concentrations of B and C the treatment is identical, even if the respective production rates are added with a term similar to eq 1.26, for example, for non-zero B_0 one has $P_B = B_0 \delta(t) + k_1 A$.

Notice that (1) eqs 1.35–c were obtained without solving any differential equations and that (2) for B_0 and $C_0 = 0$ in eqs 1.10a–c), the two sets of equations, obtained with the two methods, are identical.

1.3.2.2   REVERSIBLE FIRST-ORDER REACTIONS

$A \xleftrightarrow{} B$ (rate constants: k_1 and k_{-1})
  The -production responses and the P production rates are:

$$A_\delta(t) = \exp(-k_1 t) \tag{1.36}$$

$$B_\delta(t) = \exp(-k_{-1} t) \tag{1.37}$$

$$P_A = A_0 \delta(t) + k_{-1} B \tag{1.38}$$

$$P_B = B_0 \delta(t) + k_1 A \tag{1.39}$$

The system of coupled integral equations, $C_i(t) = P_i(t) \times C_{i\delta}(t)$, here is,

$$A = A_0 \delta(t) \otimes \exp(-k_1 t) + k_{-1} B \otimes \exp(-k_1 t) = A_0 \exp(-k_1 t) + k_{-1} B \otimes \exp(-k_1 t) \tag{1.40}$$

$$B = B_0 \delta(t) \otimes \exp(-k_{-1} t) + k_1 A \otimes \exp(-k_{-1} t) = B_0 \exp(-k_{-1} t) + k_1 A \otimes \exp(-k_{-1} t) \tag{1.41}$$

As the time evolutions of A and B are coupled, their separation can only be achieved using the Laplace transforms (L). Reminding that $L[f \times g] = L[f] \times L[g]$, $L[\Sigma a_i f_i] = \Sigma a_i L[f_i]$, and $L[a \exp(-bt)] = a/(s + b)$, one gets,

$$L[A] = \frac{A_0}{s + k_1} + \frac{k_{-1}}{s + k_1} L[B] \tag{1.42}$$

$$L[B] = \frac{B_0}{s+k_{-1}} + \frac{k_1}{s+k_{-1}} L[A] \tag{1.43}$$

This system is solved to yield, with $k = k_1 + k_{-1}$,

$$L[A] = \frac{A_0}{k}\left(\frac{k_{-1}}{s} + \frac{k_1}{s+k}\right) + B_0 \frac{k_{-1}}{k}\left(\frac{1}{s} - \frac{1}{s+k}\right) \tag{1.44}$$

$$L[B] = \frac{B_0}{k}\left(\frac{k_1}{s} + \frac{k_{-1}}{s+k}\right) + A\frac{k_1}{k}\left(\frac{1}{s} - \frac{1}{s+k}\right) \tag{1.45}$$

After Laplace inversion, L^{-1}, of the two expressions, one finally obtains,

$$A = \frac{A_0}{k}\left[k_2 + k_1 \exp(-kt)\right] + B_0 \frac{k_2}{k}\left[1 - \exp(-kt)\right] \tag{1.46}$$

$$B = \frac{B_0}{k}\left[k_1 + k_2 \exp(-kt)\right] + A_0 \frac{k_1}{k}\left[1 - \exp(-kt)\right] \tag{1.47}$$

These two equations were already obtained with the matrix eigenvalue method,[8] and have been obtained without solving any differential equation.

1.3.2.3   KINETICS IN OPEN SYSTEMS

To study these open systems with convolution kinetics, the input flow of reactants from the outside is incorporated in the production terms, whereas output flow of both reactants and products affects only the δ-responses. Consider a constant volume ideal CSTR where the following reaction occurs:

$$A \xrightarrow{k} B$$

At $t < 0$ no A is present in the reactor, then at $t = 0$ a flow of A solution, with concentration A_0 enters and leaves the reactor (which has a constant volume V), at constant volume rate. At these conditions, assuming instantaneous mixing, we have,

$$A_\delta(t) = \exp\{-(k + \Phi/V)t\} \tag{1.48}$$

$$P_A = A_o \Phi / V \tag{1.49}$$

$$B_\delta(t) = \exp(-\Phi t / V) \tag{1.50}$$

$$P_B = kA \tag{1.51}$$

Hence, reminding the fundamental relation, $C_i(t) = P_i(t) \times C_{i\delta}(t)$, we have

$$A = A_0 \frac{\Phi}{V} \otimes \exp\left\{-\left(k + \frac{\Phi}{V}\right)t\right\} \tag{1.52}$$

$$B = kA \otimes \exp\left(-\frac{\Phi}{V}t\right) = kA_0 \frac{\Phi}{V} \otimes \exp\left\{-\left(k + \frac{\Phi}{V}\right)t\right\} \otimes \exp\left(-\frac{\Phi}{V}t\right) \tag{1.53}$$

Finally one obtains,

$$A = \frac{A_0}{\dfrac{V}{\Phi}k + 1}\left\{1 - \exp\left[-\left(k + \frac{\Phi}{V}\right)t\right]\right\} \tag{1.53a}$$

$$B = \frac{kA_0}{\dfrac{V}{\Phi}k + 1}\left\{\frac{V}{\Phi}\left[1 - \exp\left(-\frac{\Phi}{V}t\right)\right] - \frac{1}{k}\left[\exp\left(-\frac{\Phi}{V}t\right) - \exp\left(-\left(k + \frac{\Phi}{V}\right)t\right)\right]\right\} \tag{1.53b}$$

Note that non-zero steady-state concentrations of A and B are attained for long times, when $B/A = kV/\Phi$.

1.4 CONCLUSION

Matrix and convolution methods, two ignored subjects in physical chemistry and in many chemical kinetics textbooks, find several applications in solving, first-order chemical and photochemical reactions. The given method constructs rate matrices of any order and complexity reduces at the level of a recipe the task of writing the K matrix, a tedious and error-prone procedure. Matrix approach allows a general view of the integration of rate equations with no direct integration, thanks to the eigenvalue method for first-order reactions and to the Euler and the more complex Runge-Kutta methods. The reader has surely noticed the similarity between the eigenvalue method for solving first-order K matrices, and the same method that is used in quantum chemistry.

The convolution approach allows writing the rate equations directly in the integrated form, whenever the decay of a species is effectively first order. The presented examples are kinetic schemes, whose results are well known, and that the reader, a bit versed in chemical kinetics, can easily check. Two additional and more complex cases could be studied with the convolution formalism: (1) reactions with time-dependent rate coefficients, including excimer formation and radiationless energy transfer and (2) reactions where the species that disappear through first-order processes belong to a kinetic system that contains a mixture of unimolecular and bimolecular steps.[16,17,21]

KEYWORDS

- **chemical kinetics**
- **matrix kinetics**
- **Eigenvalue method**
- **Euler numerical method**
- **convolution kinetics**

REFERENCES

1. Atkins, P. W. *Physical Chemistry*. Oxford University Press: Oxford, 1990.
2. Laidler, K. J. *Chemical Kinetics*. Harper & Row: New York, 1987.
3. Zsabó, Z. G. *Comprehensive Chemical Kinetics*; Bamford, C. H. et al., Eds, Elsevier: New York, Vol. 2, 1969.
4. Papula, L. *Mathematik für Chemiker*. F. Enke Verlag: Stuttgart, 1975.
5. Moore, W. J.; Pearson, R. G. *Kinetics and Mechanism*. Wiley: New York, 1981.
6. Eyring, H.; Lin, S. H.; Lin, S. M. *Basic Chemical Kinetics*. Wiley: New York, 1980.
7. Berberan-Santos, M. N.; Martinho, J. M. G. The Integration of Kinetic Rate Equations by Matrix Methods. *J. Chem. Educ.* **1990**, *67*, 375–379.
8. Pogliani, L.; Terenzi, M. Matrix Formulation of Chemical Reaction Rates. A Mathematical Chemical Exercise. *J. Chem. Educ.* **1992**, *69*, 278–280.

9. Pogliani L. How to Construct First-order Kinetic Matrices and Higher-order Kinetic Matrices. *React. Kinet. Catal. Lett.* **1993,** *49,* 345–351.
10. Pogliani L. How to Construct Kinetic Matrices: The Autocatalytic Case. *React. Kinet. Catal. Lett.* **1995, 55,** 41–46.
11. Pogliani, L. The Detailed Balance Principle in Matrix Kinetics. *React. Kinet. Catal. Lett.* **1998,** *64,* 9–14.
12. Pogliani, L. Pattern Recognition and Alternative Physical Chemistry Methodologies. *J. Chem. Inf. Comput. Sci.* **1998,** *38,* 130–143.
13. Trinajstić, N. *Chemical Graph Theory,* 2nd ed.; CRC Press: Boca Raton, 1992.
14. Kvasnička, V. Formal First-order Chemical Kinetics. *Chem. Papers* **1987,** *41,* 145–169.
15. Temkin, O. N.; Zeigarnik, A. V.; Bonchev D. G. *Chemical Reaction Networks: A Graph-theoretical Approach*; CRC Press: Boca Raton, 1996.
16. Martinho J. M. G.; Winnik M. A. Transient Effects in Pyrene Monomer-excimer Kinetics. *J. Phys. Chem.* **1987,** *91,* 3640–3644.
17. Berberan-Santos, M. N.; Martinho, J. M. G. Kinetics of Sequential Energy-transfer Processes. *J. Phys. Chem.* **1990,** *94,* 5847–5849.
18. Berberan-Santos, M. N. The Time Dependence of Rate Coefficients and Fluorescence Anisotropy for Non-delta Production. *J. Luminescence* **1991,** *50,* 83–87.
19. Berberan-Santos, M. N.; Martinho, J. M. G. A Linear Response Approach to Kinetics With Time-dependent Rate Coefficients. *Chem. Phys.* **1992,** *164,* 259–269.
20. Berberan-Santos, M. N.; Pogliani, L.; Martinho, J. M. G. A Convolution Approach to the Kinetics of Chemical and Photochemical Reactions. *React. Kinet. Catal. Lett.* **1995,** *54* 287–292.
21. Pogliani, L.; Berberan-Santos, M. N.; Martinho, J. M. G. Matrix and Convolution Methods in Chemical Kinetics. *J. Math. Chem.* **1996,** *20,* 196–210.
22. Berberan-Santos, M. N.; Farinha J. P. S.; Martinho J. M. G. Linear and Convolution Methods for the Analysis of Ground and Excited State Kinetics. Application to the Monomer-excimer Scheme. *Chem. Phys.* **2000,** *260,* 401–414.
23. Perez-Benito, J. F.; Driss Lamrhari, D.; Arias C. Three Rate Constants from a Single Kinetic Experiment: Formation, Decomposition, and Reactivity of the Chromium(VI) - Glutathione Thioester Intermediate. *J. Phys. Chem.* **1994,** *98,* 12621–12629.
24. Sauder, J. M.; MacKenzie, N. E.; Roder H. Kinetic Mechanism of Folding and Unfolding of *Rhodobacter capsulatus* Cytochrome *c*2. *Biochemistry* **1996,** *35,* 16852–16862.
25. Park, S.-Ho; O'Neil, K. T.; Roder H. An Early Intermediate in the Folding Reaction of the B1 Domain of Protein G Contains a Native-like Core. *Biochemistry* **1997,** *36,* 14277–14283.
26. Khalil, M. I. Calculating Enthalpy of Reaction by a Matrix Method. *J. Chem. Educ.* **2000,** *77,* 185–187.
27. Duffey, G. H. Phenomenological Chemical Kinetics. In *Modern Physical Chemistry*; Kluwer Academic/Plenum Publ.: New York, 2000; pp 405–433.
28. Justi, R. Teaching and Learning Chemical Kinetics. In *Chemical Education: Towards Research-based Practice;* J. K. Gilbert et al., Eds; Kluwer Academic Pub.: New York, 2002; pp 293–315.

29. Roder, H.; Maki, K.; Latypov, R. F.; Cheng, H.; Shastry, M. C. R. Early Events in Protein Folding Explored by Rapid Mixing Methods. In *Protein Folding Handbook*; J. Buchner, J.; Kiefhaber, T., Eds, Wiley-VCH Verlag: Weinheim, 2005; pp 491–535.

30. Roder, H.; Maki, K.; Cheng, H. Early Events in Protein Folding Explored by Rapid Mixing Methods. *Chem. Rev.* **2006**, *106*, 1836–1861.

31. Konermann, L.; Messinger, J.; Hillier, W. Mass Spectrometry-based Methods for Studying Kinetics and Dynamics in Biological Systems. In *Biophysical Techniques in Photosynthesis II*; Aartsma, T. J., Matysik, J., Eds.; Springer: Dordrecht, 2008; pp 167–190.

32. Harris, S. J.; Murdock, D.; Zhang, Y.; Oliver, T. A. A.; Grubb, M. P.; Orr-Ewing, A. J.; Greetham, G. M.; Clark, I. P.; Towrie, M.; Bradforth, S. E.; Ashfold, M. N. R. Comparing Molecular Photofragmentation Dynamics in the Gas and Liquid Phases. *Phys. Chem. Chem. Phys.* **2013**, *15*, 6567–6582.

33. Harris, S. J.; Murdock, D.; Grubb, M. P.; Clark, I. P.; Greetham, G. M.; Towrie, M.; Ashfold, M. N. R. Tracking a Paterno-buchi Reaction in Real Time Using Transient Electronic and Vibrational Spectroscopies. *J. Phys. Chem. A* **2014**, *118*, 10240–10245.

34. Maafi, M.; Brown, R. G. Photophysics and Kinetics of Naphthopyran Derivatives, Part 1: General Analytical Solutions for the Kinetics of AB(k,φ) and ABC(k,φ) Systems. *Int. J. Chem. Kinet.* **2005**, *37*, 162–174.

35. Kotenev, V. A.; Tyurin, D. N.; Tsivadze, A. Yu; Kiselev, M. R.; Vysotskii, V. V.; Zolotarevskii, V. I. Transformation of Metal-oxide Nanostructures in the Process of Afteroxidation of Iron Reactively Evaporated in Oxygen Atmosphere. *Protect. Metals Phys. Chem. Surf.* **2009**, *45*, 704–708.

36. Korobov V. I. ; Ochkov, V. F. *Chemical Kinetics with Mathcad and Maple*. Springer-Verlag: Wien, 2011.

37. Crouch, S. R.; Cullen, T. F.; Scheeline, A.; Kirkor, E. S. Kinetic Determinations and Some Kinetic Aspects of Analytical Chemistry. *Anal. Chem.* **1998**, *70*, 53R–106R.

38. Rae, M.; Berberan-Santos, M. N. Pre-equilibrium Approximation in Chemical And Photophysical Kinetics. *Chem. Phys.* **2002**, *280*, 283–293.

39. Holland, J. P.; Giansiracusa, J-H.; Bell, S. G.; Wong, L. -L; Dilworth, J. R. In vitro Kinetic Studies on the Mechanism Oxygen-dependent Cellular Uptake of Copper Radiopharmaceuticals. *Phys. Med. Biol.* **2009**, *54*, 2103–2119.

40. Gilmore, R. L. Laboratory Studies in Chemically Mediated Phosphorus Removal. Theses and Dissertations (Comprehensive), 2009, pp I–XIV. ISBN: 978-0-494-54228-6.

41. Berberan-Santos, M. N. Mathematical Basis of the Integral Formalism of Chemical Kinetics. Compact Representation of the General Solution of the First-order Linear Differential Equation. *J. Math. Chem.* **2010**, *47*, 1184–1188.

42. Berberan-Santos, M. N. Extending the Convolution Method: A General Integral Formalism for Chemical Kinetics. Application to Enzymatic Reactions. *MATCH Commun. Math. Comput. Chem.* **2010**, *63*, 603–622.

43. Hasan, N. A. S. B.; Balasubramanian, P. Exact Solution for the Kinetic Equations of First Order Reversible Reaction Systems through Flow Graph Theory Approach. *Ind. Eng. Chem. Res.* **2013**, *52*, 10594–10600.

44. Freitas, A. F.; Maçanita, A. A. L.; Quina, F. H. Improved Analysis of Excited State Proton Transfer Kinetics by the Combination Of Standard and Convolution Methods. *Photochem. Photobiol. Sci.* **2013,** *12,* 902–910.

45. Shammas, S. L.; Crabtree, M. D.; Dahal, L.; Wicky, B. I. M.; Clarke, J. Insights into Coupled Folding and Binding Mechanisms from Kinetic Studies. *J. Biol. Chem.* **2016,** *291,* 6689–6695.

46. Kyurkchiev, N.; Markov, S. On the Numerical Solution of the General Kinetic "K-Angle" Reaction System. *J. Math. Chem.* **2016,** *54,* 792–805.

47. Roland Tóbiás, R.; Tasi, G. Simple Algebraic Solutions to the Kinetic Problems of Triangle, Quadrangle and Pentangle Reactions. *J. Math. Chem.* **2016,** *54,* 85–99.

48. Shah, R.; Ohashi, T.; Erickson. H. P.; Oas. T. G. Spontaneous Unfolding-refolding of Fibronectin Type III Domains Assayed by Thiol Exchange. *J. Biol. Chem.* **2017,** *292* (3), 955–966.

49. Tóbiás, R.; Stacho, L. L.; Gyula Tasi, G. First-order Chemical Reaction Networks I: Theoretical Considerations. *J. Math. Chem.* **2016,** *54,*1863–1878.

50. El Seoud, O. A.; Baader,W. J.; Bastos, E. L. Practical Chemical Kinetics in Solution. In *Encyclopedia of Physical Organic Chemistry,* 1st ed.; Zerong Wang, Z., Ed.; John Wiley & Sons, Inc.: New York, 2017.

51. Farlow, S. J. *Partial Differential Equations for Scientists and Engineers.* Wiley: New York, 1982.

52. Margenau, H.; Murphy, G. M. *The Mathematics of Physics and Chemistry.* Van Nostrand: London, 1968.

CHAPTER 2

ANALYSIS OF CHEMICAL AND BIOLOGICAL KINETIC REACTOR SYSTEMS TO REMOVE HEXAVALENT CHROMIUM VIA A SULFATE-REDUCING PROCESS: MODELING AND CONTROL APPROACH

PABLO A. LÓPEZ-PÉREZ[1,*], RICARDO AGUILAR-LÓPEZ[2], and V. PEÑA-CABALLERO[3,*]

[1]Escuela Superior de Apan, Universidad Autónoma de Hidalgo, Carretera Apan-Calpulalpan, km. 8. Chimalpa Tlalayote s/n Colonia Chimalpa, Apan, 43900 Hgo. México

[2]Departamento de Biotecnología y Bioingenierнa, CINVESTAV-IPN, Av. Instituto Politécnico Nacional No. 2508, Col. San Pedro Zacatenco, 07360, Mexico City, CDMX., Mexico

[3]Departamento de Ingeniería Agroindustrial, División de Ciencias de la Salud e Ingenierías. Campus Celaya-Salvatierra, Universidad de Guanajuato, Ave. Ing. Barros Sierra No. 201, Esq. Ave. Baja California. Ejido de Santa María del Refugio, C.P. 38140, Mexico

*Corresponding author. E-mail: vicente.caballero@ugto.mx; sasve1991@yahoo.com.mx

ABSTRACT

This work focuses on the design of different closed-loop control schemes based on nonlinear controller. A simulated two-stage reactor system (TSRS) was used to remove sulfate and hexavalent chromium (Cr(VI)).

Desulfovibrio alaskensis 6SR cells were anerobically grown in the first-stage reactor (biological rector BR), and then feeding to a second-stage reactor (chemical reactor, CR) where chemically Cr(VI) reduction occurred with biosulfide. In the last 20 years, it has been demonstrated that complete Cr(VI) removal was achieved in the CR and sulfate removal from 3500 to 500 mg L^{-1} and was achieved in BR and lactate consumption from 5000 to almost 8000 mg L^{-1} under appropriate conditions. To illustrate the influence of the consumed sulfate in order to produced sulfide by *D. alaskensis* 6SR in the BR and Cr(VI) reduction with biosulfide in the CR, a mathematical model was developedt by incorporating lactate oxidation and sulfide production kinetic and Cr(VI) reduction into the mass balance relationship of BR and CR, respectively, for an RS. Analysis of lactate, sulfate, and Cr(VI) concentration in SR using both model simulation and experimental data indicated that the rate of sulfide production is function of lactate/sulfate relation and that the Cr(VI) reduction decreased with the depletion of the sulfide concentration in BR. Finally, a nonlinear controller was proposed in order to regulate the dynamic behavior of the BR for biosulfide production and regulate the dynamic behavior of the CR for Cr(VI) reduction. Numerical experiments were carried out in order to show the satisfactory performance of the proposed controller.

2.1   INTRODUCTION

Heavy metal contamination in industrial effluent becomes an important issue due to its irreversible effect on environment and human health.[1] Heavy metals can accumulate in vital organs of human and animal bodies and harmfully affect their health.[2] Some metals include manganese, nickel, cadmium, mercury, chromium, cobalt, zinc, iron, silver, and copper. Considering the harmful effects that heavy metals exert on the environment, their removal from sewage or, at least, the reduction in their concentration to the limits allowed by current strict regulations must be accomplished.[3] Several methods can be used to remove heavy metals from industrial wastewaters, including chemical precipitation, ion exchange, flotation, electrochemical techniques, membrane-related processes, and biological processes.[4,5]

Chemical methods for metal removal have a limited effectiveness, the need to use expensive chemical reagents, as well as the disposal problems of secondary wastes are the main disadvantages when the metal concentration target is very low. Bioprecipitation of metals with hydrogen sulfide produced by sulfate-reducing bacteria (SRB) has been proposed as an alternative process using sulfidogenic bioreactors, is considered to be a better technique, due to its simplicity and high efficiency, characteristics that have received much attention in recent years,[5,6,7,8] for example, the removal of metals from wastewater through the production of biogenic sulfides, followed by metal bioprecipitation.[9] The genus *Desulfovibrio* show a high efficiency in heavy metal removal in a range from 0 ppm (mg L^{-1}) to 150 mg L^{-1}, Cd^{2+} (30–150 mg L^{-1}), Cr^{6+} (75–80 mg L^{-1}), Pb^{2+} (75–80 mg L^{-1}), Zn^{2+} (30–150 mg L^{-1}).[10] Also, the SRB can produce extracellular polymeric substances (EPSs), usually comprising a mixture of polysaccharides, mucopolysaccharides, and proteins, depending on the strain and culture conditions.[11] EPSs play important roles in aggregation of bacterial cells in flocs and biofilms, stabilization of biofilm structure, retention of water, and formation of a protective barrier that buffers harmful environmental effects, but also plays a role in biosorption of heavy metals.[12] Biosorption is a fast and reversible process for the removal of toxic metal ions from wastewater by live or dried biomass, which resembles adsorption and in some cases ion exchange.[8] The biosorption offers an alternative to the remediation of industrial effluents as well as the recovery of metals contained in other media. The mechanisms by which metal ions bind onto the cell surface most likely include electrostatic interactions, van der Waals forces, covalent bonding, redox interactions, and extracellular precipitation, or some combination of these processes.[14,15]

The mechanism of the sulfate reduction for removal of sulfate, sulfide production, and heavy metals is illustrated by reactions (i–iii):

$$SO_4^{2-} + 8e^- + 4H_2O \rightarrow S^{2-} + 8OH^- \tag{i}$$

$$\underset{\text{sulfide}}{S^{2-}} + \underset{\text{heavy metal}|\text{soluble}|}{M^{2+}} \rightarrow \underset{\text{heavy metal}|\text{insoluble}|}{MS \downarrow} \tag{ii}$$

$$\underset{\text{heavy metal }|\text{insoluble}|}{MS^- \downarrow} + \underset{\text{adsorbent }|\text{biofilm}|}{B^+} \rightarrow \underset{\text{heavy metal}|\text{bioadsorption}|}{MSB} \tag{iii}$$

However, the determination of kinetic parameters throughout structured model on basis of biomass components, such as concentration of metabolites, enzymes, DNA, and/or RNA is a complex task.[16] For this reason, the kinetic parameters more commonly used are estimated through unstructured kinetic model that use measured of biomass, substrate, product, as well as yield coefficients determined in the bulk of the reactor.[17] Few kinetic models have obtained satisfactory fitting of sulfate reducing kinetic, in most cases Monod model is used. In other cases, combinations of two models were done: Haldane–Boulton, Haldane–Levenspiel, Haldane–Luong, Moser–Boulton, Moser–Levespiel, and Moser-Luong.[18,19] Moosa et al. (2002) reported the effects of sulfate concentration and temperature on bacterial growth rate. Their experimental data were fitted with different mathematical models including those of Monod, Chen and Hashimoto, and Contoins,[21] these models do not take into account the product inhibition phenomenon generated by sulfide accumulation inside bioreactor; much less the dynamic removal of Cr(VI) and Cd^{2+} by bioprecipitation and biofilms.[22] Therefore, more widely applicable models are not available, for example, to describe the combined effects of sulfate, sulfide, biomass, lactate, biofilm, acetate on the removal of Cr(VI) and Cd^{2+} in anerobic process.

Chromium has been designated as a priority pollutant of soil and ground water by the US Environmental Protection Agency. Industrial applications are the main cause of chromium contaminations. The most stable forms of chromium are chromate [Cr(VI)], which is a water-soluble, toxic, mutagenic, and carcinogenic environmental pollutant; and trivalent chromium [Cr(III)], which is less soluble, less toxic, and less mobile.[26]

Conventional waste management options for Cr(VI) removal include chemical reduction to Cr(III) followed by precipitation under alkaline conditions or removal by ion-exchange, adsorption, and membrane separation.[23,24] Among the main disadvantages of these methods are high-operating costs, need for large quantities of chemical adsorbents, requirement for preliminary treatment steps, and difficulty in treating the solid wastes thus generated.[25]

Recently, the interest in bioremediation processes has greatly increased. The two main processes under investigation are: the adsorption of Cr(VI) onto microbial cells and the reduction of Cr(VI) to Cr(III) by enzymatic reaction or indirectly by reducing compounds produced by microorganisms.[27,28] The biological reduction of hexavalent chromium has attracted increased interest, since this process may not only relieve the toxicity of chromium that affect living organisms but also may aid in the precipitation of chromium at near-neutral pH (mainly as $Cr(OH)_3$) for further physical removal. This process has also been considered as an economically feasible alternative for wastewater treatment.[27]

A wide variety of microbial species have been reported to reduce the highly soluble and toxic hexavalent chromium to the less soluble and less toxic trivalent chromium under aerobic and/or anerobic conditions.[27,28] SRB constitute a group of anerobic prokaryotes, commonly found in contaminated environments by heavy metals, metalloids, or other pollutants that are lethal to other microorganisms. SRB are able to couple the oxidation of organic compounds or hydrogen with the reduction of sulfate. It has been proposed that SRB are able to detoxify contaminated environments by an indirect chemical reduction of heavy metal via the production of H_2S, which is the end product of the dissimilatory sulfate reduction. The schematic representation of the proposed reactor system is specified in Figure 2.1.

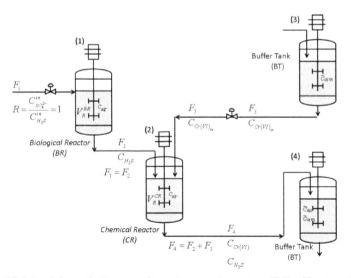

FIGURE 2.1 Schematic diagram of two-stage reactor system (RS) without control loops: sulfate was reduced by *D. alaskensis* 6SR in biological reactor BR (tank 1) and the effluent from the BR was continuously mixed with the effluent contaminated with hexavalent chromium from a holding tank (tank 3) in a chemical reactor CR, (tank 2) in order to chromium reduction with biosulfide from BR, the effluent of CR was collected in other holding tank (tank 4).

To develop cost-effective and environmentally friendly techniques to remove Cr(VI) ions from industrial wastewater, microorganisms capable of tolerating and reducing high Cr(VI) concentrations must be isolated.[29] Furthermore, the design and operation of biological processes for Cr(VI) reduction requires a thorough understanding of the effect of Cr(VI) on the kinetic characteristics of the Cr(VI) reduction processes and the mathematical model (e.g., phenomenological model) for the system biology, however, this type of information is usually scarce. Finally, the goal of this paper is to propose a novel controller algorithm to regulate the dynamic behavior of the BR for biosulfide production and regulate the dynamic behavior of the CR for Cr(VI) reduction.

2.2 EXPERIMENTAL

The bacterium *Desulfovibrio alaskensis* 6SR was used.[31] The analytic methods presented were based on a previous work of Peña-Caballero et al. (2015). The procedures of inoculation and medium have been described elsewhere.[1,33,34]

2.2.1 MODEL DEVELOPMENT

2.2.1.1 GENERAL CHARACTERISTICS OF MODELS

Consider a biological system described by a set of state variables, that is, the concentrations of substrates, metabolic products, and biomass in the extracellular solution.[36] In a batch bioreactor, the process dynamics can be described by the following model:

$$\frac{dx(t)}{dt} = \lambda \vartheta\big(x(t)\big) \tag{2.1}$$

Here,

$x \in \Re_+^n$ = states-vector of concentration, $\lambda \in \Re_+^{n \times m}$ = corresponding stoichiometric coefficients matrix (m < n), $\vartheta_i \in \Re_+^n$ = vector of reaction rates kinetic term, $t \in \Re_+$ = time.

$$\vartheta\left(x\left(t\right)\right) = \mu\left(t\right) x\left(t\right) \tag{2.2}$$

$\mu(t)$ is the specific growth rate, which is a function of the concentrations: sulfate, sulfide, biomass, lactate, acetate, metal. Such reaction rates vary with time and are usually influenced by many physicochemical and biological environmental factors like substrate, biomass and product concentrations, as well as pH, temperature, dissolved oxygen concentration, or various microbial growth inhibitors. Actually there exists no systematic rule in order to determine the best model. Moreover, the identification of their biological parameters can be time consuming. Indeed, these structures are nonlinear and, in most cases, nonlinearizable.

2.2.2 PROCESS MODEL AND PROBLEM STATEMENT

The bioprocess of sulfide production by *D. alaskensis* 6SR with chemical reduction of hexavalent chromium by sulfide (biosulfide) namely here as two-stage reactor system (TSRS). The TSRS model was nonlinear, mechanistic mathematical model, which describe the cell growth, and the cell metabolism for BR and chromium reduction in CR. It was based on the certain standard assumptions, such as well-mixed bioreactor, and the perfect control of culture temperature and pH.

The model consisted of seven equations first-order ordinary differential equations (eqs 2.3–2.9). The state variables included the biomass concentration ($C_X(t)$), lactate concentration ($C_{Lac}(t)$), sulfate concentration $\left(C_{SO_4^{2-}}(t)\right)$, sulfide concentration ($C_{H_2S}(t)$), and acetate concentration ($C_{Acet}(t)$) for BR; chromium concentration ($C_{Cr(VI)}(t)$) and sulfide concentration $\left(C'_{H_2S}(t)\right)$ for QR. The first five state variables were dependent on growth $\mu(\cdot)$, as mentioned above. $\mu(\cdot)$ refer to the specific rate of cell growth which can be a function of the concentration of any reactant or product of the stoichiometry of the reaction, that is, $C_X(t)$, $C_{Lac}(t)$, $C_{SO_4^{2-}}(t)$, $CH_2S(t)$ and $C_{Acet}(t)$. And death rate $k_d = k_4$ of the bioprocess sulfate-reducing process (eqs 2.3–2.9). In addition, the dynamics of the two final state variables are a function of the reaction rate of oxide reduction.

Biomass in liquid balance ($C_X(t)$):

$$\frac{dC_X}{dt} = k_5 \left[\frac{C_{SO_4^{2-}}}{k_2 + C_{SO_4^{2-}}} \right] \left[k_3 - C_{H_2S} \right] C_x - k_4 C_X - DC_X \tag{2.3}$$

Lactate in liquid mass balance ($C_{Lac}(t)$):

$$\frac{dC_{Lac}}{dt} = -k_6 \left[\frac{C_{SO_4^{2-}}}{k_2 + C_{SO_4^{2-}}} \right] \left[k_3 - C_{H_2S} \right] C_x + D \left[C_{Lac_{in}} - C_{Lac} \right] \tag{2.4}$$

Sulfate in liquid mass balance $\left(C_{SO_4^{2-}}(t) \right)$:

$$\frac{dC_{SO_4^{2-}}}{dt} = -k_7 \left[\frac{C_{SO_4^{2-}}}{k_2 + C_{SO_4^{2-}}} \right] \left[k_3 - C_{H_2S} \right] C_x + D \left[C_{SO_4^{2-}{}_{in}} - C_{SO_4^{2-}} \right] \tag{2.5}$$

Biosulfide in liquid mass balance ($C_{H_2S}(t)$):

$$\frac{dC_{H_2S}}{dt} = k_8 \left[\frac{C_{SO_4^{2-}}}{k_2 + C_{SO_4^{2-}}} \right] \left[k_3 - C_{H_2S} \right] C_x - D \tag{2.6}$$

Acetate in liquid mass balance ($C_{Acet}(t)$):

$$\frac{dC_{Acet}}{dt} = k_9 \left[\frac{C_{SO_4^{2-}}}{k_2 + C_{SO_4^{2-}}} \right] \left[k_3 - C_{H_2S} \right] C_x - DC_{Acet} \tag{2.7}$$

Chromium in liquid mass balance ($C_{Cr(VI)}(t)$):

$$\frac{dC_{Cr(VI)}}{dt} = -k_{10} C_{Cr(VI)} \left[C'_{H_2S} \right]^{\gamma_1} + Q \left[C_{Cr(VI)_{in}} - C_{Cr(VI)} \right] - DC_{Cr(VI)} \tag{2.8}$$

Sulfide in liquid mass balance ($C'_{H_2S}(t)$):

$$\frac{dC'_{H_2S}}{dt} = -k_{11} \left[C_{Cr(VI)} \right]^{\gamma_2} C'_{H_2S} - QC_{Cr(VI)} + D \left[C_{H_2S} - C'_{H_2S} \right] \tag{2.9}$$

where k_2 is the substrate saturation (mg L^{-1}); k_3 is growth inhibition constant by sulfide (mg L^{-1}); k_5 is the ration; k_1/k_3 (L h^{-1} mg^{-1}) with k_1 denotes the maximum specific growth rate h^{-1}; k_6 is the ration k_5/k_3. F_{BR} is the bioreactor inlet flow rate and F_{QR} is the chemical reactor inlet flow rate. F_{BR} is considered as manipulated variable. V_{BR} and V_{QR} are the reactor volume in bioreactor and chemical reactor, respectively. $D = F_1/V_{BR}$; $Q = F_3/V_{CR}$ are parameterized variables (Fig. 2.1). And with the initial conditions at $t = 0$; $C_X(0)$, $C_{Lac}(0)$, $C_{SO_4^{2-}}(0)$, $C_{H_2S}(0)$, $C_{Acet}(0)$, $C_{Cr(VI)}$, C'_{H_2S}. The mass balances for the bioreactor (eqs 2.1–2.7) have implicated the kinetics of cell growth through the specific speed of cell growth in eq 2.10.

$$r_{C_x} = \mu\left(C_{SO_4^{2-}}, C_{H_2S}\right) = k_5 \left[\frac{C_{SO_4^{2-}}}{k_2 + C_{SO_4^{2-}}}\right]\left[k_3 - C_{H_2S}\right]C_x \left\{C_{H_2S} < k_3\right. \quad (1.10)$$

$$\left(\frac{\text{cell mass produced}}{\text{unit volume} \times \text{time}}\right)$$

Model in eq 1.10 assumes a critical inhibitor (i.e., for product inhibition biosulfide) concentration above which cells cannot grow.

2.2.3 SYSTEM STABILITY

The system was stable considering the Lyapunov criterion on a point of equilibrium of the process $(\bar{x})$ because the determinant of matrix A was different from zero, that is, its eigenvalues (λ_i; $i = 1,2, \ldots, n$, $n = 7$) were negative ($\lambda_i < 0$), for which we considered the linearized version of the model in eqs 2.3–2.9:

$$\dot{x} = Ax + Bu$$
$$y = Cx + Du \quad\quad (2.11)$$

where　$\bar{x} = \left[\bar{C}_X, \bar{C}_{Lac}, \bar{C}_{SO_4^{2-}}, \bar{C}_{H_2S}, \bar{C}_{Acet}, \bar{C}_{Cr(VI)}, \bar{C}_{H_2S}\right]^T$;　$A = \left.\frac{\partial f}{\partial x}\right|_{\bar{x}}$, $B = \left.\frac{\partial f}{\partial u}\right|_{\bar{x}}$, $D = 0, det(A) \neq 0$, $f = \left[f_1(\cdot), f_2(\cdot), f_3(\cdot), f_4(\cdot), f_5(\cdot), f_6(\cdot), f_7(\cdot)\right]^T$ are vector of equilibrium states, Jacobian matrix, control matrix, output matrix, output control matrix, determinant of matrix A, and matrix of functions, respectively.

2.2.4 PARAMETRIC IDENTIFICATION OF THE MODEL

The model parameters were determined either experimentally of fitted using the libraries of the MatLab® software. Table 2.1 presents the model parameters for this system. This model was used within the process simulator and within nonlinear control framework. In this context, the following assumptions are made for the development of a controller for TSRS (eqs 2.3–2.9) (Fig. 2.2).

Assumption 1: It is assumed that the reactor system can be properly described by eqs 2.3–2.9.

Assumption 2: A measurement of sulfate concentration $C_{SO_4^{2-}}$ sulfide concentration C_{H_2S}, C'_{H_2S}, and chromium concentration $C_{Cr(VI)}$, are assumed available on line.

Assumption 3: State variables are bounded and positive. This can be assumed because these variables represent to concentration in the bioreactor.

TABLE 2.1 Values of Parameters Describing Reactor System: Sulfate-reducing Process with *D. alaskensis* 6SR. Units According to Expressions (2.3–2.9).

Parameters	Value
k_1	0.106 h^{-1}
k_2	873.49 mgL^{-1}
k_3	627.350 mgL^{-1}
k_4	0.0058 mgL^{-1}
k_5	0.000169 Lh^{-1} mg^{-1}
k_6	(0.0024) 14.3141 gg^{-1}
k_7	(0.0021) 12.4655 gg^{-1}
k_8	(0.000286) 1.6969 gg^{-1}
k_9	(0.0015) 8.972 gg^{-1}
k_{10}	$1.6 \dfrac{L^{-\gamma_1} h^{-\gamma_1}}{mg^{-\gamma_1}}$
k_{11}	$3.6 \dfrac{L^{-\gamma_2} h^{-\gamma_2}}{mg^{-\gamma_2}}$
y_1	1.6
y_2	3.9

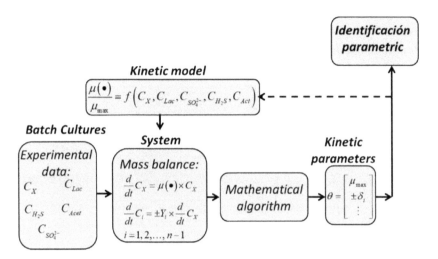

FIGURE 2.2 Parametric estimation architecture.

2.2.5 NONLINEAR CONTROL DESIGN

The control strategies on the TSRS were configured in the MatLab® environment. The nonlinear controller was used to control the sulfate and chromium in BR and QR, respectively, using the nonlinear dynamic model of TSRS (eqs 2.3–2.9).

Now, let us consider a generalized space state representation of the system 2.3–2.9 as the differential system defined on domain :

$$\begin{cases} \dot{x} = F\big(x(t), u(t)\big) \\ \quad y = h\big(x(t)\big) \\ x(0) = x_0, u \in \Omega \end{cases} \qquad (2.12)$$

$x(t) \in \mathfrak{R}^n, u(t) \in \mathfrak{R}^m, m \leq n, y(t) \in \mathfrak{R}^p \, p < n$.

Consider differential system (2.9) where $f\big(x(t), u(t)\big) = f(x) + g(x)u$ that is:

$$\begin{cases} \dot{x} = f(x) + g(x)u \\ \quad y = h(x) \\ x(0) = x_0, u \in \Omega \end{cases} \qquad (2.13)$$

where $f(\cdot)$ and $g(\cdot)$ are both smooth functions $\mathfrak{R}^n \to \mathfrak{R}^n$ and the observation function $h(\cdot)$ is also smooth $\mathfrak{R}^n \to \mathfrak{R}^p$.

Now consider the set $\Psi \in \mathfrak{R}^7_+$ as the corresponding physically realizable domain, such that:

$$\Psi = \Big\{ \big(C_X, C_{Lac}, C_{SO_4^{2-}}, C_{H_2S}, C_{Acet}, C_{Cr(VI)}, C'_{H_2S} \big) \in \mathfrak{R}^7_+ \;/\; 0 \le C_X \le C_{X max} \;;\; 0 \le C_{Lac}$$

$$\le C_{Lac\,max} \;;\; 0 \le C_{SO_4^{2-}} \le C_{SO_4^{2-}\,max} \;;\; 0 \le C_{H_2S} \le C_{H_2S max} \;;\; 0 \le C_{Cr(VI)}$$

$$\le C_{Acet\,max} \;;\; 0 \le C_{Acet} \le C_{Cr(VI)max} \;;\; 0 \le C'_{H_2S} \le C'_{H_2S max} \Big\}; f(x) \in C^\infty; f(0)$$

$$= 0 \text{ and } \|f(x)\| \le L, \|g(x)\| \le G; \forall x \in \mathfrak{R}^7_+, L, G < \infty.$$

where $x = \Big[C_X, C_{Lac}, C_{SO_4^{2-}}, C_{H_2S}, C_{Acet}, C_{Cr(VI)}, C'_{H_2S} \Big]^T$. Now consider the three cases for system control:

$$h(x) = C_{SO_4^{2-}}, \; g(x) = \Big[0, 0, C_{SO_4^{2-}\,in}, 0, 0, 0 \Big]^T,$$

$$h(x) = C_{Cr(VI)}, \; g(x) = \Big[0, 0, 0, 0, C_{Cr(VI)in}, 0 \Big]^T, \text{ and}$$

with bout cases $h(x) = C_{SO_4^{2-}}, h(x) = C_{Cr(VI)}, g(x) = \Big[0, 0, C_{SO_4^{2-}\,in}, 0, C_{Cr(VI)in}, 0 \Big]^T$.

When $u = D$.

$$f(x) = \begin{bmatrix} k_5 \left[\dfrac{C_{SO_4^{2-}}}{k_2 + C_{SO_4^{2-}}} \right] [k_3 - C_{H_2S}] C_x - k_4 C_X - u C_X \\[3mm] -k_6 \left[\dfrac{C_{SO_4^{2-}}}{k_2 + C_{SO_4^{2-}}} \right] [k_3 - C_{H_2S}] C_x + u [C_{Lac\,in} - C_{Lac}] \\[3mm] -k_7 \left[\dfrac{C_{SO_4^{2-}}}{k_2 + C_{SO_4^{2-}}} \right] [k_3 - C_{H_2S}] C_x - u C_{SO_4^{2-}} \\[3mm] k_8 \left[\dfrac{C_{SO_4^{2-}}}{k_2 + C_{SO_4^{2-}}} \right] [k_3 - C_{H_2S}] C_x - u C_{H_2S} \\[3mm] -k_{10} C_{Cr(VI)} [C'_{H_2S}]^{\alpha_1} + Q [C_{Cr(VI)in} - C_{Cr(VI)}] \\[3mm] -k_{11} [C_{Cq21\|1r(VI)}]^{\alpha_2} C_{H_2S} + Q [C_{H_2S} - C'_{H_2S}] \end{bmatrix}$$

2.2.5.1 THE PROPOSAL

In this section, $\|\cdot\|$ denotes norm of vector or matrix and $(\cdot)^T$ denotes transpose of vector or matrix.

Proposition 1. The following control input is a controller for the system 2.3–2.9:

$$D(t)_j \triangleq u(t)_j + \bar{D}_j = \alpha_{1j} \left[\frac{\beta_{1j} - \exp\left(-\beta_{2j}\left(x - x_{sp}\right)\right)}{\beta_{3j} + \exp\left(-\beta_{4j}\left(x - x_{sp}\right)\right)} \right] + \bar{D}_j ; j = 1, 2, \ldots, m, m = 3 \quad (2.14)$$

Here, in the proposed controller, sub-index j refers to its application for each regulation situation in the reactor system and $\bar{D}_j$ is a nominal feed rate to the reactor.

$$\left\{ u(x(t))_j = \sup \alpha_{1j} \left[\frac{\beta_{1j} - \exp\left(-\beta_{2j}e\right)}{\beta_{3j} + \exp\left(-\beta_{4j}e\right)} \right] \leq \alpha_{1j} \left[\frac{\beta_{1j} - \exp\left(-\beta_{2j}e\right)}{\beta_{3j} + \exp\left(-\beta_{4j}e\right)} \right] \leq \alpha_{1j} \right\} \quad (2.15)$$

The regulation error is defined as

$$e = x - x_{sp}$$

Inspired by Pablo et al. (2010), the dynamics of the controller on the system was stable, with the proposal of the following Lyapunov function

$$V = e^T Q e = \|e\|_Q^2, Q = Q^T > 0 \quad (2.16)$$

The time derivative along the trajectories of eq 2.16 is

$$\dot{V} = \dot{e}^T Q e + e^T Q \dot{e} \quad (2.17)$$

and

$$\dot{V} \leq 2\left[L + G\alpha_1\right]\|e\|_Q \quad (2.18)$$

If chosen, $a_1 < 0$

$$G\alpha_1 > L$$

$$\alpha_1 > -\frac{L}{G}$$

$$\dot{V} \leq 2\left[L + G\alpha_1\right]\|e\|_Q \leq 0$$

Full details of the controller stability test are described by Pablo et al. (2015).

2.2.5.2 CONTROLLER PERFORMANCE INDEX

The performance of each simulated controller was quantified by calculating integral of the time multiplied by the absolute error (ITAE). This is defined in eq 2.19.

$$ITAE = \int_0^t t \left| e(t) \right| dt \tag{2.19}$$

Where $e(t)$ is the error between process variable and the defined setpoint. ITAE index was selected because it penalizes errors that persist for a long period of time.

2.3   RESULTS AND DISCUSSIONS

The proposed system can be considered as an integral component for the treatment of wastewater contaminated with sulfates or chrome. The utilization of the hydrogen sulfide produced in the bioprocess of sulfate reduction to hydrogen sulfide by the *D. alaskensis* 6SR bacterium to reduce hexavalent chromium requires the optimization of the performance of the bioprocess by controlling the concentrations of the sulfate, sulfide, and chromium in the reactors. Therefore, it is a desirable target for control. Firstly, the modeling of the system is shown in Figures 2.3 and 2.4.

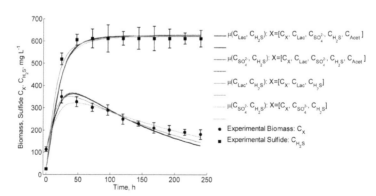

FIGURE 2.3 Experimental and model predicted concentration profiles for lactate oxidation by *D. alaskenisis* 6SR; the average time profiles of biomass and sulfide for several kinetic models.

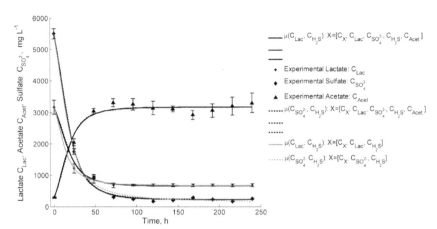

FIGURE 2.4 Experimental and model predicted concentration profiles for lactate oxidation by *D. alaskenisis* 6SR; the average time profiles of lactate, sulfide, and acetate for several kinetic models.

Different nonstructured models were tested to identify the cell growth kinetics of the bacteria $(\mu(\cdot))$ (see Fig. 2.2). The sulfate-reducing process was identified with a kinetics as a function of the concentration of sulfate and sulfide $(\mu(C_{SO_4^{2-}}, C_{H_2S}))$ for a balance of matter in the eqs 2.3–2.7. The structure of reaction kinetics was a modification of the model proposed by Hand et al. (1988). Other unstructured kinetic models were tested to identify cell growth kinetics (data not shown here), that is,

$$\mu(C_{Lac}, C_{H_2S}) / \mu_{max} = f(C_{Lac}, C_{H_2S}) : x = [C_X, C_{Lac}, C_{SO_4^{2-}}, C_{H_2S}, C_{Acet}]^T \in \mathbb{R}^{1\times5}, \mu(C_{SO_4^{2-}}, C_{H_2S}) /$$

$$?_m ax = f(C_{(SO_4^{(2-)})}, C_{(H_2S)}) : x = [C_X, C_L ac, C_{(SO_4^{(2-)})}, C_{(H_2S)}, C_A cet]^T \in \mathbb{R}^{(1\times5)}, \mu(C_{Lac}, C_{H_2S}) / \mu_{max} =$$

$$f(C_L ac, C_{(H_2S)}) : x = [C_X, C_L ac, C_{(H_2S)}]^T \in \mathbb{R}^{(1\times3)}, \mu(C_{SO_4^{2-}}, C_{H_2S}) / \mu_{max} : x = [C_X, C_{SO_4^{2-}}, C_{H_2S}]^T \in \mathbb{R}^{1\times3}$$

see Figures 2.3 and 2.4.

2.3.1 SYSTEM SIMULATION

The final result of all manipulation is the model which describes the fermentation process which is given in eqs 2.3–2.9, with initial conditions C_X (0) (time zero) = 113.4 mg L^{-1}, C_{Lac} (0) (time zero) = 5307 mg L^{-1}, $C_{SO_4^{2-}}$ (0) (time zero) = 3500 mg L^{-1}, C_{H_2S} (0) (time zero) = 24.00, C_{Acet} (0) (time zero) = 0.00 mg L^{-1}, $C_{Cr}(VI)$ (time zero) = 0.00 mg L^{-1}, C'_{H_2S}

(time zero) = 0.00 mg L^{-1} (but if $t > 0$; $0 \le C_{H_2S} \le C_{H_2S_{max}}$ mg L^{-1}), and the following constrains $C_x > 0$; $0 \le C_{Lac} < C_{Lac_{in}}$; $0 \le C_{SO_4^{2-}} \le C_{SO_4^{2-}}$; $C_{H_2S} > 0$; $C_{Acet} > 0$; $0 \le C_{Cr(VI)} \le C_{Cr(VI)_{in}}$; $0 \le C_{H_2S} \le C'_{H_2S}$.

After several dynamic simulations, tests were performed to study steady-state behavior of the reactor system model in different regions of the operating space using MatLab® software. For example, the region of the operating test corresponding to $C_{Lac_{in}} = 2000$ mg·L^{-1}; $C_{SO_4^{2-}_{in}} = 3500$mg·$L^{-1}$; $C_{Cr(VI)_{in}} = 60$mg·L^{-1}, $D = 0.025 h^{-1}$a, and $Q = 0.025 h^{-1}$ was selected as a case study to apply the proposed controller extending the model in eqs 2.3–2.9 to continuous operation. The results for these conditions to open-loop were $C_X = 81$mg·L^{-1}, $C_{Lac} = 572.6$mg·L^{-1}, $C_{SO_4^{2-}} = 2251.6$mg·L^{-1}, $C_{H_2S} = 170$mg·L^{-1}, $C_{Acet} = 895.2$mg·L^{-1}, $C_{Cr(VI)} = 1.3$mg·L^{-1}, and $C'_{H_2S} = 1.98$mg·L^{-1}. The dynamics of the states in the chemical reactor with the influent current from the BR is shown in Figure 2.5.

This steady state was stable according to the negative eigenvalues $\lambda_j; i = 1,2,…, n$, with $n = 7$: –0.012, –0.012, –0.0216, –0.042, –0.012 for BR, and – 1.049 and –3.481 for CR for system in eqs 2.3–2.9. Numerically for this operation condition, it was observed that the residual concentration of sulfate is high in the effluent in BR (>500 mg L^{-1}) and in operating the CR with the same feed rate in both reactors it can be seen that the residual concentration of chromium in the effluent stream of CR was >0.5 mg L^{-1} (Fig. 2.5), this indicates that the effluents cannot discharge into the receiving bodies as they do not comply with the regulations, then it is required to regulate the residual concentration of sulfate and chromium by controlling the flow from influent to BR and CR.

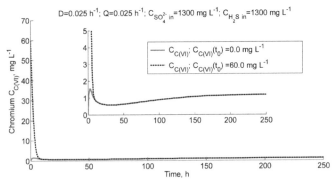

FIGURE 2.5 **(See color insert.)** Dynamic behavior of the chromium variable in operating mode in the chemical reactor (CR) in continuous: open loop for two initial conditions of hexavalent chromium in CR.

2.3.2 SYSTEM CONTROL

To regulate the sulfate and chromium concentrations in the system, the proposed controller was implemented numerically on TSRS. Inspired different control objectives on the system of reactors in eqs 2.3–2.9, as illustrative examples, were considered as three cases of regulation proposed.

Case A, regulation of the residual concentration of sulfate in the BR (Fig. 2.6); case B, regulation of the residual concentration of hexavalent chromium in the CR (Fig. 2.7); and case C, simultaneous regulation of the residual concentrations of sulfate and chromium (Fig. 2.8), respectively, BR and CR. In all the cases, the proposed controller (eq 2.15) was used to regulate the sulfate and chromium concentrations.

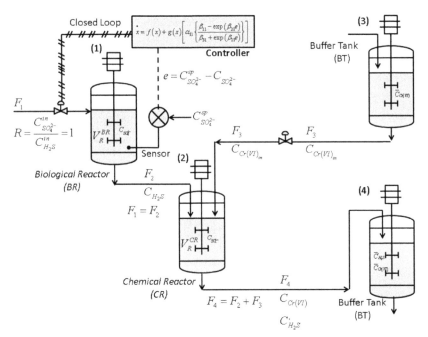

FIGURE 2.6 Reactor system for removing sulfates and chromium with control in BR in order to regulate sulfate concentration in BR.

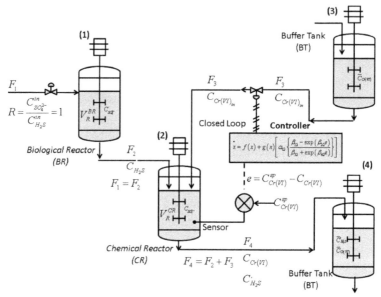

FIGURE 2.7   Reactor system for removing sulfates and chromium with control in QR in order to regulate chromium concentration in QR.

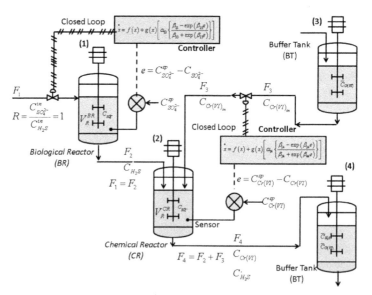

FIGURE 2.8   Reactor system for removing sulfates and chromium with both control in BR and QR in order to regulate sulfate concentration and chromium concentration in BR and QR, respectively.

2.3.3   CASES A AND B: REGULATION OF THE SULFATE CONCENTRATION AND CHROMIUM CONCENTRATION ON BR AND CR

The performance of the controller proposes for each case has been individually evaluated by simulation for step changes for the dilution rate in BR.

The system in eqs 2.3–2.9 was integrated using software MatLab®. Thus, the estate responses x, in closed loop, due to input u and initial conditions is shown in Figures 2.9 and 2.10 (red line) when the reference for sulfate concentration in BR was evaluated for following setpoints:

$$C_{SO_4^{2-} stp} \Rightarrow \begin{cases} set\ point1 & 50 \le t \le 100 \\ set\ point2 & 100 \le t \le 150 \\ set\ point3 & 150 \le t \le 250 \end{cases} \qquad (2.20)$$

It can be observed in the Figures 2.9 and 2.10 that the dynamics of the open-loop states are the continuous lines. Regulating the inflow feed flow with Q, the residual chromium concentration in the CR was regulated for different reference values for chromium concentration in the same intervals of time like (eq 2.20) Figures 2.11 and 2.12. The control of the biological process by manipulating the inlet flow (D) to the reactor to regulate the sulfate concentration was carried out until reaching a residual concentration of 500 mg L^{-1} in the BR (Fig. 2.9, red line). It can be seen that the performance of the controller is adequate to regulate the sulfate residual concentration. The response of the uncontrolled variable (zero dynamics) corresponding to the sulfur is stable (Fig. 2.10). The control of the residual concentration of chromium in the chemical reactor is shown in Figure 2.11. For different setpoint concentrations, and likewise, the sulfide response is stable. Open-loop states are distinguished by continuous lines. These results show that it is required to simultaneously regulate the concentration of sulfate and chromium in the biological and chemical reactors, respectively, to achieve maximum sulfur production for the reduction of chromium.

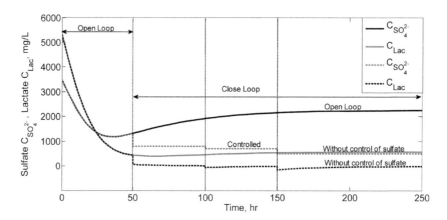

FIGURE 2.9   (See color insert.) Dynamic behavior of the state variables in operating mode in the bioreactor (BR) in continuous: open-loop black solid line and gray solid line for sulfate and lactate, respectively, and close-loop black dashed line and red dashed line for sulfate and lactate, respectively.

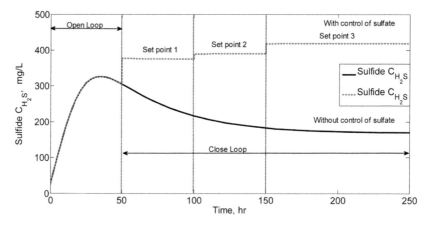

FIGURE 2.10   (See color insert.) Dynamic behavior of the sulfide variable in operating mode in the bioreactor (BR) in continuous: open-loop black solid line and close-loop red dashed line.

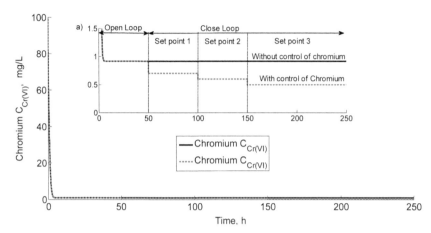

FIGURE 2.11 (**See color insert.**) Residual chromium concentration.

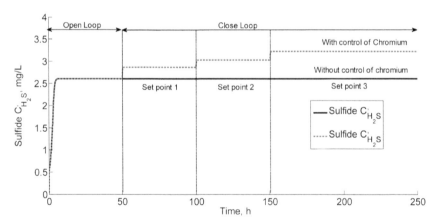

FIGURE 2.12 (**See color insert.**) Dynamic behavior of the sulfide variable in operating mode in the bioreactor (CR) in continuous: open-loop black solid line and close-loop red dashed line.

2.3.4 CASES C SIMULTANEOUS REGULATION OF SULFATE AND CHROMIUM CONCENTRATIONS IN BR AND CR, RESPECTIVELY

For this case of numerical analysis, the regulation of the sulfate residual concentration to 0 mg/L in the RB is considered for a production of hydrogen sulfide higher than 400 mg/L but lower than the concentration of inhibition (627 mg/L), thus the proposed controller was applied to regulate the concentration of sulfate in the BR to produce more sulfide to react with Cr(VI) in CR and simultaneously the controller was applied to regulate the concentration of chromium in CR for different setpoints (Tables 2.1 and 2.2).

TABLE 2.2   Values of Setpoint for Open-Loop and Closed-Loop for the BR.

Setpoint path	Setpoint open-loop	Setpoint closed-loop	Final value (mgL⁻¹)		
			Biological reactor (BR)		
			C_{Lac}	$C_{SO_4^{2-}}$	C_{H_2S}
1–3	$0 \leq t \leq 250$	-	2251.6	572.6	170.0
2–4	-	$50 \leq t \leq 100$	802.6	431.6	375.2
4–5	-	$100 \leq t \leq 150$	702.5	4.6	389.6
5–6	-	$150 \leq t \leq 250$	502.6	0	418.4

TABLE 2.3   Values of Residual Chromium and Sulfide in Cr in Setpoint for Open-Loop and Closed-Loop.

Setpoint	Setpoint open-loop (h)	Setpoint closed-loop (h)	Final value (mgL⁻¹)	
			Chemical reactor (CR)	
			$C_{Cr(VI)}$	C'_{H_2S}
-	$0 \leq t \leq 250$	-	100.00–0.9	0.00–2.6
1	-	$50 \leq t \leq 100$	0.9–0.7	2.6–2.9
2	-	$100 \leq t \leq 150$	0.7–0.6	2.9–3.0
3	-	$150 \leq t \leq 250$	0.6–0.5	3.0–3.2

As can be observed in Figure 2.13, the proposed methodology acts almost immediately, leading the chromium trajectory to the corresponding setpoints without overshoots and settling times, with an adequate effort of the control as shown in Figure 2.14. Note that all the controllers act over reachable operating regions, which is feasible for a real implementation. Figure 2.15 shows the closed-loop behavior of the uncontrolled sulfide concentration; as can be observed, it is stable trajectories; therefore, it can be concluded that the TSRS is stabilizable. The simulated TSRS performances for different setpoints (eq 2.20) were quantified using ITAE (eq 2.19) (Table 2.4). The TSRS $\left(C_{Lac_{in}} / C_{SO_4^{2-}_{in}} = 0.5 \right)$ illustrated is optimality in setpoint tracking which resulted in the lowest performance index.

TABLE 2.4 Simulated TSRS Performance Index for TSRS.

Index	TSRS $\left(C_{Lac_{in}} / C_{SO_4^{2-}_{in}} \right)$ 0.5	TSRS $\left(C_{Lac_{in}} / C_{SO_4^{2-}_{in}} \right)$ 0.75	TSRS $\left(C_{Lac_{in}} / C_{SO_4^{2-}_{in}} \right)$ 1.5
ITAE	53,100	54,250	84,210

$C_{Lac_{in}} = 2000 mg \cdot L^{-1}; C_{SO_4^{2-}_{in}}$

Figure 2.16 is related to a 3D phase portrait, including all the trajectories of the state variables. Note that the trajectories remain in a bounded region. In this phase, plane shows the trajectories of the variables in open-loop (black line) and in closed-loop (dashed red line) according to the setpoint in Tables 2.2 and 2.3.

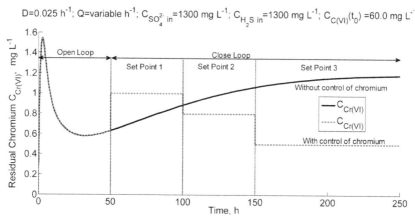

FIGURE 2.13 (See color insert.) Dynamic behavior of the chromium variable in operating mode in the chemical reactor (CR) in continuous: open-loop black solid line and close-loop red dashed line.

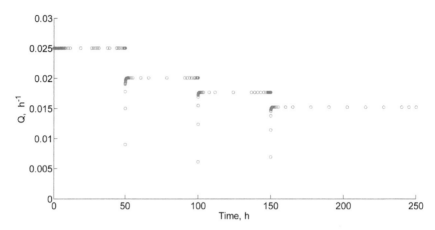

FIGURE 2.14   **(See color insert.)** Control input efforts.

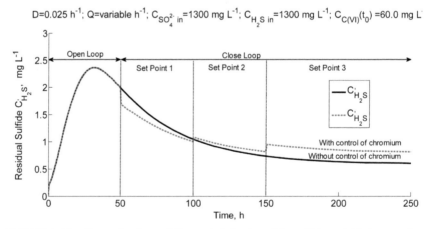

FIGURE 2.15   **(See color insert.)** Dynamic behavior of the sulfide variable in operating mode in the chemical reactor (CR) in continuous: open-loop black solid line and close-loop red dashed line.

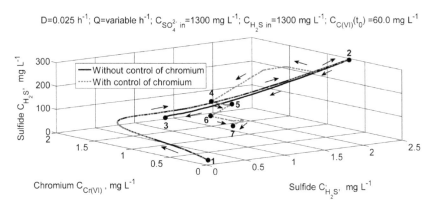

FIGURE 2.16 **(See color insert.)** Open- loop and closed-loop phase portrait.

The tuning of the proposed controller for the three case studies is summarized in Table 2.5.

TABLE 2.5 Values of the Parameters of the Controller in eq 2.14 for the Three Case Studies to Regulate the TSRS.

Case	Case A	Case B	Case C	
Parameters	$\alpha_{11} = 1$	$\alpha_{12} = 10\ h^{-1}$	$\alpha_{13} = \alpha_{11}$	$\alpha_{14} = \alpha_{12}$
	$\beta_{11} = 0.1$	$\beta_{12} = 1$	$\beta_{13} = \beta_{11}$	$\beta_{14} = \beta_{12}$
	$\beta_{21} = 1$	$\beta_{22} = 1\ mg\ L^{-1}$	$\beta_{23} = \beta_{21}$	$\beta_{24} = \beta_{22}$
	$\beta_{31} = 1$	$\beta_{32} = 0.1$	$\beta_{33} = \beta_{31}$	$\beta_{34} = \beta_{32}$
	$\beta_{41} = 1$	$\beta_{42} = 1\ L\ mg^{-1}$	$\beta_{43} = \beta_{41}$	$\beta_{44} = \beta_{42}$

2.4 CONCLUSIONS

In this paper, a class of nonlinear controller with tangent hyperbolic output injection which provides stabilization under the considered assumptions is proposed. The proposed controller is able to provide adequate performance for regulation and tracking purposes. Moreover, no perfect knowledge of the model is required to design the controller which is a valuable feature process application. Finally, biological and chemical reactor system to remove heavy metals as hybrid reactor systems design will be of the particular interest to conducting research in the fields of the nonlinear control systems.

KEYWORDS

- sulfate-reducing process,
- *Desulfovibrio alaskensis* 6SR
- chromium reduction
- nonlinear control
- biosulfide

REFERENCES

1. Hernández Melchor, D. J.; López-Pérez, P. A.; Carrillo Vargas, S.; Alberto Murrieta, A.; González Gómez, E.; Camacho-Perez, B. Experimental and Kinetic Study for Lead Removal via Photosynthetic Consortia Using Genetic Algorithms to Parameter Estimation. *Environ. Sci. Pollution Res.* **2007**.
2. Dua, M.; Singh, A.; Sethunathan, N.; Johri, A. K Biotechnology and Bioremediation: Successes and Limitations. *Appl. Microbiol. Biot.* **2002**, *59*, 143–152.
3. Sud, D.; Mahajan, G.; Kaur, M. P. Agricultural Waste Material as Potential Adsorbent for Sequestering Heavy Metal Ions from Aqueous Solutions. *Biores. Technol.* **2008**, *99*, 6017–6027.
4. Doana, H. D.; Saidi, M. Simultaneous Removal of Metal Ions and Linear Alkylbenzene Sulfonate by Combined Electrochemical and Photocatalytic Process. *J. Hazardous Mater.* **2008**, *158*, 557–567.
5. Nadeem, M.; Shabbir, M.; Abdullah, M. A.; Shah, S. S.; McKay, G. Sorption of Cadmium from Aqueous Solution by Surfactant-modified Carbon Adsorbents. *Chem. Eng. J.* **2009**, *148*, 365–370.
6. Tomei, F. A.; Barton, L. L.; Lemanski, C. L.; Zocco, T. G.; Fink, N. H.; Sillerud, L. O. Transformation of Selenate and Selenite to Elemental Selenium by Desulfovibrio Desulfuricans. *J. Ind. Microbiol.* **1995**, *14*, 329–336.
7. Benner, S. G.; Blowes, D. W.; Ptacek, C. J.; Mayer, K. U. Rates of Sulfate Reduction and Metal Sulfide Precipitation in a Permeable Reactive Barrier. *Appl. Geochem.* **2002**, *17* (3), 301–320.
8. Volesky B. Detoxification of Metal-bearing Effluents: Biosorption for the Next Century. *Hydrometallurgy* **2001**, *59*, 203–216.
9. Muyzer, G.;Stams, A. J. M. The Ecology and Biotechnology of Sulfate-reducing Bacteria. *Nat. Rev. Microbiol.* **2008**, *6*, 441–454.
10. Kieu, H. T. Q.; Müller, E.; Horn, H. Heavy Metal Removal in Anaerobic Semi-continuous Stirred Tank Reactors by a Consortium of Sulfate-reducing Bacteria. *Water Res.* **2011**, *45* (13), 3863–38670. doi:10.1016/j.watres.2011.04.043.
11. Castro, H. F.; Williams, N. H.; Andrew, O. Phylogeny of Sulfate-reducing Bacteria. *FEMS Microbiol. Ecol.* **2000**, *31*, 1–9.

12. Beech, I. B.; Sunner, J. Biocorrosion: Towards Understanding Interactions Between Biofilms and Metals. *Curr. Opin. Biotechnol.* **2004,** *15,* 181–186.

13. Volesky, B. Detoxification of Metal-bearing Effluents: Biosorption for the Next Century. *Hydrometallurgy* **2001,** *59,* 203–216.

14. Blanco, A. Immobilization of Non-viable Cyanobacteria and Their Use for Heavy Metal Adsorption from Water. In *Environmental Biotechnology,* Olguin, E. J.; Sánchez, G.; Hernández, E., Eds, Philadelphia, 2000, 135–151.

15. Gavrilescu, M. Removal of Heavy Metals from the Environment by Biosorption. *Eng. Life Sci.* **2004,** *4,* 219–232.

16. Stitt, M.; Sulpice R.; Keurentjes, J. 2010. Metabolic Networks: How to Identify Key Components in the Regulation of Metabolism and Growth. *Plant Physiol.* **2010,** *152,* 428–444.

17. Villadsen, J.; Nielsen J.; Lidén, G. 2011. *Growth Kinetics of Cell Cultures. Bioreactor Engineering Principles.* Springer, 2011.

18. Monod, J. The Growth of Bacterial Cultures. *Ann. Rev. Microbiol.* **1949,** *3,* 371–394.

19. Hidaka, T.; Horie, T.; Akao, S.; Tsuno, H. Kinetic Model of Thermophilic L-Lactate Fermentation by Bacillus Coagulants Combined with Real-time PCR Quantification. *Water Res.* **2010,** *44,* 2554–2562.

20. Moosa, S.; Nemati, M.; Harrison, S. T. L. A Kinetic Study on Anaerobic Reduction of Sulfate, Part I, Effect of Sulfate Concentration. *Chem. Eng. Sci.* **2002,** *57,* 2773–2780.

21. Moosa, S.; Nemati, M.; Harrison, S. T. L. A Kinetic Study on Anaerobic Reduction of Sulfate, Part II, Incorporation of Temperature Effects in the Kinetic Model. *Chem. Eng. Sci.* **2005,** *60,* 3517–3524.

22. Okpokwasili, G. C.; Nweke, C. O. Microbial Growth and Substrate Utilization Kinetics. *Afr. J. Biotechnol.* **2005,** *5,* 305–317.

23. Stasinakis, A. S.; Thomaidis, N. S.; Mamais, D.; Karivali, M.; Lekkas, T. D. Chromium Species Behavior in the Activated Sludge Process. *Chemosphere* **2003,** *52,* pp 1059–1067.

24. Sumathi, K. M. S.; Mahimairaja, S.; Naidu, R., Use of Low-cost Biological Wastes and Vermiculite for Removal of Chromium from Tannery Effluent. *Biores. Technol.* **2005,** *96* (3), 309.

25. Saxena, S.; Srivastava, R. K.; Samaddar, A. B. Sustainable Waste Management Issues in India. *IUP J. Soil Water Sci.* **2010,** *3* (1), 72–90.

26. Cheung, K. H.; Gu, J. –D. Mechanism of Hexavalent Chromium Detoxification by Microorganisms and Bioremediation Application Potential: A Review. *Int. Biodeterioration Biodegradation* **2007,** *59* (1), 8–15.

27. Cheung, K. H.; Gu, J. D. Reduction of Chromate ($CrO42-$) by an Enrichment Consortium and an Isolate of Marine Sulfate-reducing Bacteria. *Chemosphere* **2003,** *52,* 1523–1529.

28. Desjardin, V.; Bayard, R.; Lejeune, P.; Gourdon, R. Utilisation of Supernatants of Pure Cultures of Streptomyces Thermocarboxydus NH50 to Reduce Chromium Toxicity and Mobility in Contaminated Soils. *Water Air Soil Pollution* **2003,** *3,* 153–160.

29. Chardin, B.; Chaspoul, F.; Gallice, P.; Bruschi, M. Chromium Speciation in Bacterial Culture Medium by Combining Strong Anion Exchange Liquid Chromatography with Inductively Coupled Plasma Mass Spectrometry. Application To the Reduction

of Cr(Vi) by Sulfate-reducing Bacteria. *J. Liquid Chromatogr. Related Technol.* **2002**, *25* (6), 877–887.

30. Peña-Caballero, V.; Aguilar-López, R.; López-Pérez, P. A.; Neria, I. Reduction of Cr(VI) Utilizing Biogenic Sulfide: An Experimental and Mathematical Modeling Approach. *Desalination Water Treatment* 2015.

31. Neria-González, I.; Wang, E. T.; Ramírez, F.; Romero, J. M.; Hernández-Rodríguez, C. Characterization of Bacterial Community Associated to Biofilms of Corroded Oil Pipelines from the Southeast of Mexico. *Anaerobe* **2006**, *12*, 122–133.

32. Postgate, J. R. *In the Sulfate-reducing Bacteria*, 2nd ed., Cambridge University Press, 1984.

33. Cord-Ruwisch, R. A Quick Method for Determination of Dissolved and Precipitated Sulfides in Cultures of Sulfate-reducing Bacteria. *J. Microbiol. Methods* **1985**, *4*, 33–36.

34. APHA-AWWA-WPCF. Standard Methods for the Examination of Water and Wastewater. American Public Health Association, American Water Works Association Water Pollution Control Federation, Washington, 1975, pp 800–869.

35. Velázquez-Sánchez, H. G.; Figueroa-Estrada, J. C.; López-Pérez, P. A.; Aguilar-López, R. *Uncertainty Observer-based I/O Linearizing Control for the Regulation of a Continuous Wastewater Bioreactor for Cd Removal. Chapter Two.* Nova Science Publishers, pp 39–70.

36. Dochain, D. State and Parameter Estimation in Chemical and Biochemical Processes: A Tutorial. *J. Proc. Control* **2003**, *13*, 801–818.

37. Perrier, M.; Dochain, D. Evaluation of Control Strategies for Anaerobic Digestion Process. *Int. J. Adapt Control Signal Process* **1993**, *7*, 309–321.

38. López-Pérez, A. P.; Aguilar-López, R. Dynamic Nonlinear Feedback for Temperature Control of Continuous Stirred Reactor with Complex Behaviour. *J. Appl. Res. Technol.* **2009**, *7*, 202–217.

39. López-Pérez, P. A.; Cuevas-Ortiz, F. A.; Rigel, V.; Gómez-Acata; Aguilar-López, R. Improving Bioethanol Production via Nonlinear Controller with Noisy Measurements. *Chem. Eng. Comm.* **2015**, *202* (11), 1438–1445.

40. López Pérez, P. A.; Neria-González, M. I.; Aguilar-López, R. Increasing the Bio-hydrogen Production in a Continuous Bioreactor Via Nonlinear Feedback Controller. *Int. J. Hydrogen Energy* **2015**, *40* (48), 17224–17230.

CHAPTER 3

FLUORESCENCE EXCITATION AND EMISSION: COMPARISION WITH ABSORPTION

FRANCISCO TORRENS[1,*] and GLORIA CASTELLANO[2]

[1]*Institut Universitari de Ciència Molecular, Universitat de València, Edifici d'Instituts de Paterna, P. O. Box 22085, E-46071 València, Spain*

[2]*Departamento de Ciencias Experimentales y Matemáticas, Facultad de Veterinaria y Ciencias Experimentales, Universidad Catolica de Valencia San Vicente Mártir, Guillem de Castro-94, E-46001 València, Spain*

*Corresponding author. E-mail: torrens@uv.es

ABSTRACT

In 20th century, analytical techniques based on the interaction of light with matter began to be used. This chapter discusses: (1) instruments for molecular spectrofluorometry (the difference between fluorometers and spectrofluorometers), (2) absorption, excitation, and emission, (3) Stokes shift and the mirror image rule, (4) how an analysis of tools of photochemistry in the study of advanced materials should include energy and electron transfer theories, studies on the interactions of dyes with deoxyribonucleic acid, nanomaterials, and sensors, (5) how people need to recognize the important symbiotic relationship between theory, application, and instrumentation in analytical science, (6) fluorescence (FLU) as an emission spectroscopy, which allows two modes: FLU emission spectrum and FLU excitation spectrum (similar to UV–visible absorption spectrum), (7) two important properties in fluorescent molecules: quantum yield and Stokes shift, and (8) a comparison of UV–visible with FLU spectroscopies.

3.1 INTRODUCTION

In earlier publications, it was informed UV–visible (UV–VIS) absorption spectroscopy, its link with fluorescence (FLU)[1] and mass spectrometry as an only method with many ionization techniques.[2] The present report deals with molecular FLU.[3–9]

3.2 BACKGROUND

Figure 3.1 shows Jablonski diagram: a partial-energy diagram for a photoluminescent system.[10] FLU is the radiative emission from the first singlet excited state (S_1) to the singlet ground state S_0. Fluorescent emission can be considered the processes opposite to excitation (absorption). However, comparing both spectra, they do not superpose but result rather mirror images (mirror image rule), finding FLU emission spectrum shifted (Stokes shift) to longer wavelengths (Stokes rule).

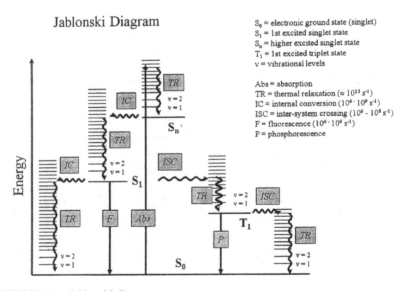

FIGURE 3.1   Jablonski diagram.

Some definitions follow.

K_F is the probability by unit of time that the electron return to ground state by FLU emission.

$\tau_0 = \dfrac{1}{K_F}$ is the natural FLU lifetime or intrinsic lifetime of the excited state or natural FLU time.

$\tau = \dfrac{1}{K_F + K_N}$ is FLU lifetime, where: $K_N = K_{IC} + K_{ISC} + K_{Predis} + K_{Dis}$, Predis denotes pre-dissociation and Dis, dissociation.

$I_{F_t} = I_{F_0} e^{-(K_F + K_N)t} = I_{F_0} e^{-\frac{t}{\tau}}$ FLU intensity decays exponentially with time.

$\Phi_F = \dfrac{K_F}{K_F + K_N} = \dfrac{\tau}{\tau_0}$ FLU quantum yield is the ratio between the number of photons emitted by FLU and number of photons absorbed by the molecule.

3.3   QUENCHING

There are three types of FLU decay (quenching).

3.3.1   COLLISIONAL QUENCHING

Collisional quenching is because of molecular collisions. It rises with temperature and decays with viscosity coefficient.

3.3.2   STATIC QUENCHING

Static quenching is independent of molecular collisions. It depends on neither temperature nor viscosity coefficient.

3.3.2   FÖRSTER RESONANCE ENERGY TRANSFER

Förster resonance energy transfer (FRET): (1) Molecule A is excited; (2) Molecule A passes the energy to molecule B by dipolar interaction (this interaction decays with the intermolecular distance as $1/R^6$ and requires short A–B distance about 40 Å); and (3) Molecule B emits FLU. FRET is typical in biochemistry between proteins. It allows to identify protein complexes and pairs of proteins are in contact in these complexes.

3.4   FACTORS AFFECTING MOLECULAR FLU

Factors affecting molecular FLU are: (1) presence of chromophore groups; (2) chemical structure; (3) steric factors; and (4) physicochemical parameters: temperature, viscosity coefficient, and pH.

3.5   FLUORESCENT MARKERS

Not all organic compounds emit FLU (native FLU) but other molecules are obtained capable of developing FLU from non-FLU compounds via their binding (FLU markers, fluorogenic reagents, or fluorophores). Fluorogenic reaction should be versatile and fast, not demanding drastic conditions and obtained products should not easily degrade. A fluoroinductor molecule with native FLU should present the following characteristics: (1) FLU emission is different from reaction milieu; (2) high FLU quantum yield; and (3) fluorogen and reaction product must be stable in the reaction milieu. Most used FLU markers are: fluorescein, different rhodamines, orthophthaldehyde (OPA), fluorescein isothiocyanate (FITC), fluorescamine, chlorophyll, chlortetracycline, and so on.

3.6   INSTRUMENTS FOR MOLECULAR SPECTROFLUOROMETRY

The fundamental characteristic of instruments for fluorometric measures is the incorporation of two selectors of radiations, situated before and after sample cell and arranged generally in an angle of 90°. Two types of instruments exist.

3.6.1   FLUOROMETERS

Selector of radiation is a filter. In general, they do not incorporate a plotter. Radiation source is $Hg_{(v)}$ lamp.

3.6.2 SPECTROFLUOROMETERS

Selector of radiation is a diffraction net. They incorporate a plotter. Radiation source is Xe arc lamp.

3.7 APPLICATIONS

Molecular FLU presents the following applications: (1) qualitative and quantitative titration of substance with native FLU [e.g., quinine sulphate, group-B vitamins (Vits), etc.]; (2) qualitative and quantitative titration of substances that develop FLU after a chemical reaction; (3) qualitative and quantitative titration of substances marked with FLU compounds (e.g., immunofluorescence); and (4) it is a chosen technique for titrating: (1) rye-ergot and rauvolfia alkaloids, (2) actinomycin-A and griseofulvin, (3) anphetamines, (4) barbiturics, (5) Lysergsäure-Diethylamid (LSD), morphine, and heroin, being FLU able to differentiate all of them, (6) dibenzodiazepines, (7) organo-phosphorated and -chlorated pesticides, and (8) some analgesics, group-B Vits and cardiotonic glycosides (digitoxin and digoxin).

3.8 PHYSICAL CHEMICAL PRINCIPLES OF PHOTOVOLTAICS: SOLAR CELLS

Bisquert group reviewed the status of the understanding of dye-sensitized solar cells (DSSC), emphasizing clear physical models with predictive power, and discussed them in terms of the chemical and electrical potential distributions in the device.[11] Before doing so, they placed DSSC in the overall picture of photovoltaic (PV) energy converters, reiterating the fundamental common basis of all PV systems and their most important differences.

3.9 ANALYSIS OF TOOLS OF PHOTOCHEMISTRY IN ADVANCED-MATERIALS STUDY

An analysis of tools of photochemistry in the study of advanced materials should include: (1) energy and electron transfer theories; (2) studies on the interactions of dyes with deoxyribonucleic acid (DNA); (3) nanomaterials; and (4) sensors. Some applications of FLU include: unusually strong delayed FLU of C_{70},[12] thermally activated delayed FLU as a cycling process between excited singlet and triplet states, and application to the fullerenes,[13] oxygen-proof FLU temperature sensing with pristine C_{70} encapsulated in polymer nanoparticles (NPs),[14] optical sensing and imaging of trace oxygen with record response,[15] sensing and imaging of oxygen with parts per billion limits of detection, and based on the quenching of the delayed FLU of $^{13}C_{70}$ fullerene in polymer hosts,[16] and dual FLU sensor for trace oxygen and temperature, with unmatched range and sensitivity.[17] In this laboratory, González-Béjar group ultraclean derivatized monodisperse Au NPs via light amplification by stimulated emission of radiation (laser) drop ablation customization of polymorph Au nanostructures (cf. Fig. 3.2).[18]

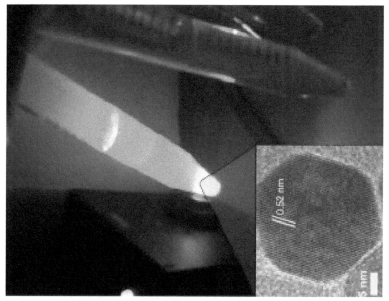

FIGURE 3.2   Nanoparticles activated by light.
Source: Used with permission from Ref. [20].

They informed cucurbit[n]uril-capped up-conversion NPs as highly emissive scaffolds for energy acceptors (cf. Fig. 3.3).[19]

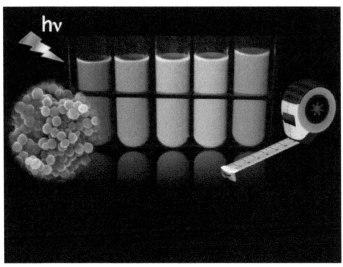

FIGURE 3.3 The dark side of the nanoworld.
Source: Used from R. E. Galian.

They reported up-conversion NPs for bioimaging and regenerative medicine (cf. Fig. 3.4).[20]

FIGURE 3.4 Nanoworld.
Source: Used with permission from R. E. Galian.

3.10 CRITICAL APPROACH TO PSEUDOSCIENCES AS DIDACTIC EXERCISE

Gaita-Ariño group proposed three teaching activities.[21] They ranged in complexity, and were adequate for implementation in classrooms and laboratories from secondary education to university level. The leitmotiv for the activities were series dilutions, and the concepts of substance quantity, mol and concentration. Transversally, this was used to point out the pseudoscientific fraud of homeopathy, which in recent years was alarmingly popular. An important benefit of a basic understanding of chemical concepts was the ability to detect the kind of hocus-pocus. One of the problems dealt with how many dilutions one should do to reach an only Au NP. Taking into account the rise experienced by nanoscience and nanotechnology field, the fact of working with pupils the synthesis and characterization of NPs of interest in a number of fields became a tool useful from the motivational aspect. For that, a simple synthesis procedure of Au NPs was incorporated. Au NPs can be related to experiences in informal contexts.

3.11 DISCUSSION

People need to recognize the important symbiotic relationship between theory, application, and instrumentation in analytical science.[22] Three views of scientific inquiry exist:

1. Classical view:

   Theory $\xrightarrow{\text{Guides}}$ Application

2. The converse side:

   Application $\xrightarrow{\text{Drives}}$ Theory

3. Instrumentation provides a catalyst:

   Application $\xleftarrow{\text{Instrumentation}}$ Theory

FLU is an emission spectroscopy, which allows two modes: (1) FLU emission spectrum and (2) FLU excitation spectrum (similar to UV–VIS absorption spectrum). Two properties are important in molecular FLU: high quantum yield and high Stokes shift.

On comparing FLU with UV–VIS spectroscopies, FLU presents advantages: (1) it is more sensitive and can be applied to lower sample concentrations (10^{-10} M in FLU vs. 10^{-5} M in UV–VIS), (2) greater selectivity (in FLU, two parameters are controlled vs. only one in UV–VIS), and (3) lower bands superposition in mixtures because of the lower number of FLU bands. On the other hand, UV–VIS presents the advantage of its generality as analytical technique: (1) UV–VIS can be applied to a greater number of molecules than FLU and (2) in UV–VIS, linear calibration extends to greater concentrations than in FLU.

3.12   FINAL REMARKS

From the present discussion, the following final remarks can be drawn.

1. An analysis of tools of photochemistry in the study of advanced materials should include: energy and electron transfer theories, studies on the interactions of dyes with deoxyribonucleic acid, nanomaterials and sensors. Some applications of FLU were discussed.
2. People need to recognize the important symbiotic relationship between theory, application and instrumentation in analytical science. Three views of scientific inquiry exist: theory guides application, application drives theory, and instrumentation provides a catalyst for both application and theory.
3. FLU is an emission spectroscopy, which allows two modes: FLU emission spectrum and FLU excitation spectrum (similar to UV–VIS absorption spectrum).
4. Two properties are important in fluorescent molecules: high quantum yield and high Stokes shift.
5. A comparison of FLU with UV–VIS spectroscopies showed that both analytical techniques are complementary with regard to selectivity, sensitivity, bands superposition in mixtures, and generality.

ACKNOWLEDGMENTS

The authors thank support from Generalitat Valenciana (Project No. PROMETEO/2016/094) and Universidad Católica de Valencia *San Vicente Mártir* (Project Nos. UCV.PRO.17-18.AIV.03 and 2019-217-001).

KEYWORDS

- chromophore
- electronic state
- emission spectrum
- excitation spectrum
- fluorescence lifetime
- fluorochrome
- fluorophore

REFERENCES

1. Torrens, F.; Castellano, G. "Ultraviolet-Visible Absorption Spectroscopy: Link with Fluorescence" In "Research Methods and Applications in Chemical and Biological Engineering"; Pourhashemi, A., Deka, S.C., Haghi, A.K., Eds.; Apple Academic-CRC: Waretown, NJ, 2019, pp 37–45.
2. Torrens, F.; Castellano, G. "Mass Spectrometry: The Only Method with Many Ionization Techniques" In "Research Methods and Applications in Chemical and Biological Engineering"; Pourhashemi, A., Deka, S.C., Haghi, A.K., Eds.; Apple Academic-CRC: Waretown, NJ, 2019, pp 47–57.
3. Skoog, D. A.; Holler, F. J.; Crouch, S. R. *Principles of Instrumental Analysis*; Cengage Learning: Boston, MA, 2018.
4. Valls Planells, O.; del Castillo García, B., Eds. *Técnicas Instrumentales en Farmacia y Ciencias de la Salud*; Universidad Norbert Wiener: Lima, Peru, 2009.
5. Robinson, J. W.; Frame, E. M. S.; Frame, G. M., II. *Undergraduate Instrumental Analysis*; CRC: Boca Raton, FL, 2014.
6. Miñones Trillo, J. *Manual de Técnicas Instrumentales*; Círculo Editor Universo: Barcelona, Spain, 1978.
7. Levine, I. N. *Molecular Spectroscopy*; Wiley: Chichester, UK, 1975.
8. Turro, N. J.; Ramamurthy, V.; Scaiano, J. C. *Modern Molecular Photochemistry of Organic Molecules*; Viva Books: Delhi, India, 2017.
9. Valeur, B.; Barberan-Santos, M. N. *Molecular Fluorescence: Principles and Applications*; Wiley-VCH: Weinheim, Germany, 2012.
10. Gierschner, J. *Book of Abstracts 2008, Optical Spectroscopy on Organic Materials, Valencia, Spain, 2008*; Universitat de València: València, Spain, 2008: O-1.
11. Bisquert, J.; Cahen, D.; Hodes, G.; Rühle, S.; Zaban, A. Physical Chemical Principles of Photovoltaic Conversion with Nanoparticulate, Mesoporous Dye-Sensitized Solar Cells. *J. Phys. Chem. B* **2004**, *108*, 8106–8118.
12. Barberan-Santos, M. N.; Garcia, J. M. M. Unusually Strong Delayed Fluorescence of C_{70}. *J. Am. Chem. Soc.* **1996**, *118*, 9391–9394.

13. Baleizão, C.; Barberan-Santos, M. N. Thermally Activated Delayed Fluorescence as a Cycling Process Between Excited Singlet and Triplet States: Application to the Fullerenes. *J. Chem. Phys.* **2007**, *126*, 204510–204518.

14. Augusto, V.; Baleizão, C.; Barberan-Santos, M. N.; Farinha, J. P. S. Oxygen-Proof Fluorescence Temperature Sensing with Pristine C_{70} Encapsulated in Polymer Nanoparticles. *J. Mater. Chem.* **2010**, *20*, 1192–1197.

15. Nagl, S.; Baleizão, C.; Borisov, S. M.; Schäferling, M.; Barberan-Santos, M. N.; Wolfbeis, O. S. Optical Sensing and Imaging of Trace Oxygen with Record Response. *Angew. Chem. Int. Ed.* **2007**, *46*, 2317–2319.

16. Kochmann, S.; Baleizão, C.; Barberan-Santos, M. N.; Wolfbeis, O. S. Sensing and Imaging of Oxygen with Parts per Billion Limits of Detection and Based on the Quenching of the Delayed Fluorescence of $^{13}C_{70}$ Fullerene in Polymer Hosts. *Anal. Chem.* **2013**, *85*, 1300–1304.

17. Baleizão, C.; Nagl, S.; Schäferling, M.; Barberan-Santos, M. N.; Wolfbeis, O. S. Dual Fluorescence Sensor for Trace Oxygen and Temperature with Unmatched Range and Sensitivity. *Anal. Chem.* **2008**, *80*, 6449–6457.

18. Bueno-Alejo, C. J.; D'Alfonso, C.; Pacioni, N. L.; González-Béjar, M.; Grenier, M.; Lanzalunga, O.; Alarcon, E. I.; Scaiano, J. C. Ultraclean Derivatized Monodisperse Gold Nanoparticles Through Laser Drop Ablation Customization of Polymorph Gold Nanostructures. *Langmuir* **2012**, *28*, 8183–8189.

19. Francés-Soriano, L.; González-Béjar, M.; Pérez-Prieto, J. Cucurbit[*n*]uril-Capped Upconversion Nanoparticles as Highly Emissive Scaffolds for Energy Acceptors. *Nanoscale* **2015**, *7*, 5140–5146.

20. González-Béjar, M.; Francés-Soriano, L.; Pérez-Prieto, J. Upconversion Nanoparticles for Bioimaging and Regenerative Medicine. *Front. Bioeng. Biotechnol.* **2016**, *4*, 47.

21. Abellán, G.; Rosaleny, L. E.; Carnicer, J.; Baldoví, J. J.; Gaita-Ariño, A. La Aproximación Crítica a las Pseudociencias como Ejercicio Didáctico: Homeopatía y Diluciones Sucesivas. *An. Quím.* **2014**, *110*, 211–217.

22. Hieftje, G. M. Editorial: The Two Three Sides of Spectroscopic Investigation. *Spectroscopist* **2018**, *2018* (2), 5.

CHAPTER 4

ETHNOBOTANICAL STUDIES OF MEDICINAL PLANTS: UNDERUTILIZED WILD EDIBLE PLANTS, FOOD, AND MEDICINE

FRANCISCO TORRENS[1,*] and GLORIA CASTELLANO[2]

[1]Institut Universitari de Ciència Molecular,
Universitat de València, Edifici d'Instituts de Paterna,
P. O. Box 22085, E-46071 València, Spain

[2]Departamento de Ciencias Experimentales y Matemáticas,
Facultad de Veterinaria y Ciencias Experimentales,
Universidad Catolica de Valencia San Vicente Mártir,
Guillem de Castro-94, E-46001 València, Spain

*Corresponding author. E-mail: torrens@uv.es

ABSTRACT

The use of plants, and traditional medicine, plays an important role for the discovery of pharmacological agents. Ethnopharmacology represents a multidimensional approach, shaped by tradition and science, which can improve people knowledge of plant use and local meaning of health and disease. Medicinal plants are an important element of indigenous medical system. In spite of the fact that the local population presents access to modern medicines, people depend, at least for the treatment of some diseases, on herbal remedies. It is important to educate local people about respecting their natural habitats, and strengthen the legislation against the illegal trade of endemic and endangered plants in the local and national market. Insights into neurobiology together with chemistry and pharmacology, however, open completely new avenues into chemical approaches toward lovesickness. For treating the addictive component and intrusive thinking associated with lovesickness, one could experiment with the food additive β-caryophyllene and the nutraceutical N-acetylcysteine. Researchers suggest conducting clinical trials on the patients, besides the conventional medical treatments, to obtain a better results via the herbs recommended by ethnomedicine. If satisfied, it is recommended to produce the herbs in the form of conventional medical drugs and be included in the pharmacopoeia of medicine. Continuity and change exist in ethnopharmacology, a transdisciplinary science for people's future. People are increasingly aware how crucial the protection and sustainable use of biodiversity is, and that an intrinsic link exists between biological and cultural diversity.

4.1   INTRODUCTION

In a philosophy in traditional Chinese medicine (TCM), plants aromatic components generally suppress microorganisms. Underutilized wild edible plants were reviewed as a source of food and medicine. Some herbs affecting vitiligo were revised based on Avicenna's *Canon of Medicine*. Herbal products regulations were discussed in a few countries. *Lagerstroemia speciosa* plant was examined with potential antidiabetic effect.

Lagerstroemia fauriei and Lophira lanceolata ethanolic extracts were recommended as substitute indicators in acid base titrimetry.[1] It was informed that the hepatoprotective potential of *Ficus auriculata* and *Sarcochlamys pulcherrima* are the two ethnomedicinal plants used by the Mishing community of Assam. Phenolic compounds were identified from nettle as new candidate inhibitors of main enzymes responsible on type-II diabetes.

Earlier publications in *Nereis*, and so on classified yams,[2] lactic acid bacteria,[3] fruits,[4] food spices,[5] and oil legumes[6] by principal component, cluster, and meta-analyses. The molecular classifications of 33 phenolic compounds derived from the cinnamic and benzoic acids from *Posidonia oceanica*,[7] 74 flavonoids,[8] 66 stilbenoids,[9] 71 triterpenoids and steroids from *Ganoderma*,[10] 17 isoflavonoids from *Dalbergia parviflora*,[11] 31 sesquiterpene lactones (STLs),[12,13] and STL artemisinin derivatives[14] were informed. A tool for interrogation of macromolecular structure was reported.[15] Mucoadhesive polymer hyaluronan favors transdermal penetration absorption of caffeine.[16,17] Polyphenolic phytochemicals in cancer prevention and therapy, bioavailability, and bioefficacy were reviewed.[18] From Asia to Mediterranean, soya bean, Spanish legumes, and commercial soya bean principal component, cluster, and meta-analyses were informed.[19] Natural antioxidants from herbs and spices improved the oxidative stability and frying performance of vegetable oils.[20] The relationship between vegetable oil composition and oxidative stability was revealed via a multifactorial approach.[21] Chemical and biological screening approaches to phytopharmaceuticals[22] and cultural interbreeding in indigenous/scientific ethnopharmacology were informed.[23] The aim of this work is to review ethnobotanical studies of medicinal plants, underutilized wild edible plants, food and medicine, bioactive compounds in *Zingiber zerumbet* rhizomes essential oils (EOs), anti-allergic and immunomodulatory properties, and the ethnopharmacology of love.

4.2 BIOACTIVES IN *Zingiber zerumbet* OILS: ANTI-ALLERGY AND IMMUNOMODULATION

Bioactive compounds in *Zingiber zerumbet* rhizomes EOs, anti-allergic, and immunomodulatory properties were reviewed.[24] The major compounds that can be found are zerumbone, pinene, humulene, linalool, caryophyllene, borneol, and limonene (cf. Fig. 4.1).

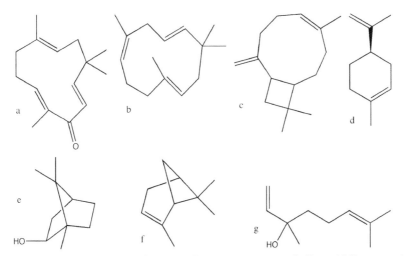

FIGURE 4.1 (a) Zerumbone; (b) α-humulene; (c) *trans*-caryophyllene; (d) limonene; (e) borneol; (f) α–pinene; and (g) linalool.

4.3   ETHNOPHARMACOLOGY OF LOVE

Ethnopharmacology of love was revised.[25]. Is there a cure for infatuation (cf. Figs. 4.2–4.4)?

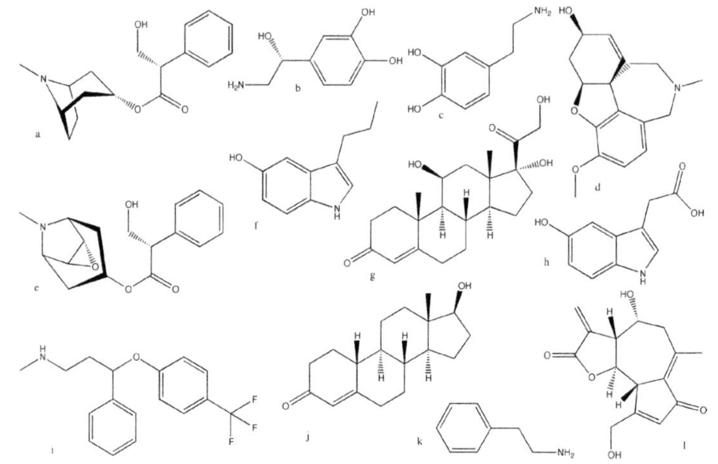

FIGURE 4.2   (a) L-Hyoscyamine; (b) noradrenalin; (c) dopamine; (d) galanthamine; (e) scopolamine; (f) serotonin; (g) cortisol; (h) 5-hydroxyindoleacetic acid; (i) fluoxetine; (j) testosterone; (k) phenylethylamine; and (l) lactucin.

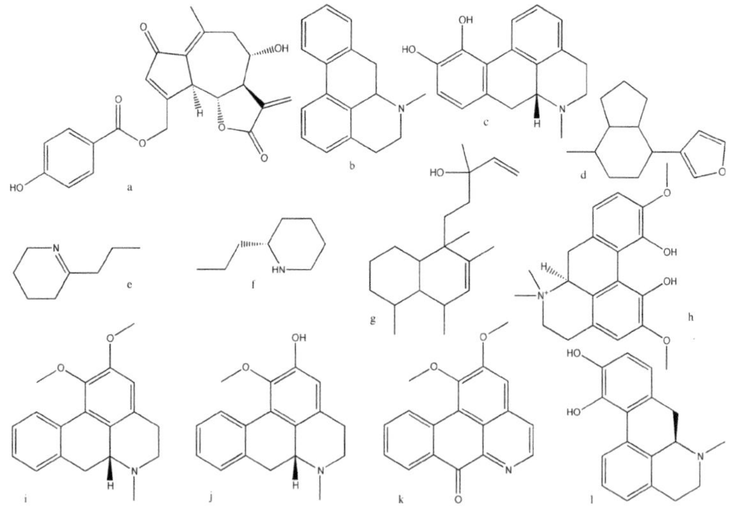

FIGURE 4.3   (a) Lactucopicrin; (b) aporphine; (c) apomorphine; (d) 5-(3-furyl)-8-methyloctahydroindolizine; (e) γ-coniceine; (f) coniine; (g) clerodadienol; (h) magnoflorine; (i) nuciferine; (j) asimilobine; (k) lysicamine; and (l) apomorphine.

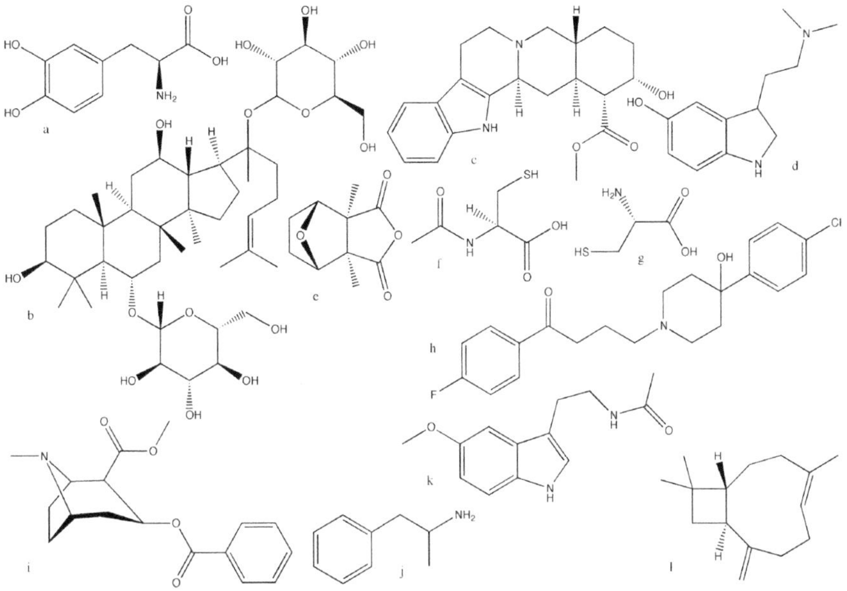

FIGURE 4.4 (a) L-DOPA; (b) ginsenoside; (c) yohimbine; (d) bufotenine; (e) cantharidin; (f) *N*-acetylcysteine; (g) cysteine; (h) haloperidol; (i) cocaine; (j) amphetamine; (k) melatonin; and (l) (*E*)-β-caryophyllene.

4.4 DISCUSSION

The use of plants, and traditional medicine, plays an important role for the discovery of pharmacological agents.[26] Ethnopharmacology represents a multidimensional approach, shaped by tradition and science, which can improve people knowledge of plant use and local meaning of health and disease. Medicinal plants are an important element of indigenous medical system. In spite of the fact that the local population presents access to modern medicines, people depend, at least for the treatment of some diseases, on herbal remedies. The therapies represent for many local doctors a low-cost alternative. The endemic species of an area present medicinal properties dependent on the metabolites, which respond to environmental stimuli that may be absent under culture conditions. Owing to this, it is important to educate local people about respecting their natural habitats, and strengthen the legislation versus the illegal trade of endemic and endangered plants in the local and national market.

The common food additive and cannabinoid receptor type-2 (CB2) agonist, β-caryophyllene, might present the potential to attenuate dopaminergic firing, quenching the reward and motivation associated with romantic love. Insights into neurobiology together with chemistry and pharmacology, however, open completely new avenues into chemical approaches toward lovesickness. For treating the addictive component and intrusive thinking associated with lovesickness, one could experiment with the food additive β-caryophyllene and the nutraceutical N-acetylcysteine.

Some herbs affecting diseases based on ethnomedicine. Researchers suggest conducting clinical trials on the patients, besides the conventional medical (CM) treatments, to obtain a better results via the herbs recommended by ethnomedicine. If satisfied, it is recommended to produce the herbs in the form of CM drugs and be included in the pharmacopoeia of medicine. Continuity and change exist in ethnopharmacology, a transdisciplinary science for people's future. People are increasingly aware on how crucial the protection and sustainable use of biodiversity is, and that an intrinsic link exists between biological and cultural diversity. Ideas taken from traditional knowledge systems result an important source of inspiration in drug development and, at the same time, new food supplements and herbal remedies from traditional societies are entering the market at an ever increasing rate.

4.5   FINAL REMARKS

From the present results and discussion the following final remarks can be drawn.

1. In spite of the fact that the local population presents access to modern medicines, people depend, at least for the treatment of some diseases, on herbal remedies. It is important to educate local people about respecting their natural habitats, and strengthen the legislation against the illegal trade of endemic and endangered plants in the local and national market.

2. Insights into neurobiology together with chemistry and pharmacology, however, open completely new avenues into chemical approaches toward lovesickness. For treating the addictive component and intrusive thinking associated with lovesickness,

one could experiment with the food additive β-caryophyllene and the nutraceutical *N*-acetylcysteine.

3. Researchers suggest conducting clinical trials on the patients, besides the conventional medical treatments, to obtain better results via the herbs recommended by ethnomedicine. If satisfied, it is recommended to produce the herbs in the form of conventional medical drugs and be included in the pharmacopoeia of medicine.

4. Continuity and change exist in ethnopharmacology, a transdisciplinary science for people's future. People are increasingly aware on how crucial the protection and sustainable use of biodiversity is, and that an intrinsic link exists between biological and cultural diversity. Ideas taken from traditional knowledge systems result an important source of inspiration in drug development.

ACKNOWLEDGMENTS

The authors thank support from Generalitat Valenciana (Project No. PROMETEO/2016/094) and Universidad Católica de Valencia *San Vicente Mártir* (Project Nos. UCV.PRO.17-18.AIV.03 and 2019-217-001).

KEYWORDS

- natural indicator
- synthetic indicator
- ethnopharmacology
- traditional medicine
- ethnobotany
- essential oil
- rhizome

REFERENCES

1. Abuh, L. O.; Shaibu, L. E.; Egu, S. A.; Omeiza, F. S. *Langerstroemia fauriei* and *Lophira lanceolata* ethanolic extracts as substitute indicators in acid base titrimetry. *J. Basic Appl. Res. Int.* **2018**, *24*, 25–30.

2. Torrens-Zaragozá, F. Molecular Categorization of Yams by Principal Component and Cluster Analyses. *Nereis* **2013**, *2013*(5), 41–51.

3. Torrens-Zaragozá, F. Classification of Lactic acid Bacteria Against Cytokine Immune Modulation. *Nereis* **2014**, *2014*(6), 27–37.

4. Torrens-Zaragozá, F. Classification of Fruits Proximate and Mineral Content: Principal Component, Cluster, Meta-Analyses. *Nereis* **2015**, *2015*(7), 39–50.

5. Torrens-Zaragozá, F. Classification of Food Spices by Proximate Content: Principal Component, Cluster, Meta-Analyses. *Nereis* **2016**, *2016*(8), 23–33.

6. Torrens, F.; Castellano, G. From Asia to Mediterranean: Soya Bean, Spanish Legumes and Commercial *Soya Bean* Principal Component, Cluster and Meta-Analyses. *J. Nutr. Food Sci.* **2014**, *4*(5), 98.

7. Castellano, G.; Tena, J.; Torrens, F. Classification of Polyphenolic Compounds by Chemical Structural Indicators and Its Relation to Antioxidant Properties of *Posidonia oceanica* (L.) Delile. *MATCH Commun. Math. Comput. Chem.* **2012**, *67*, 231–250.

8. Castellano, G.; González-Santander, J. L.; Lara, A.; Torrens, F. Classification of Flavonoid Compounds by Using Entropy of Information Theory. *Phytochemistry* **2013**, *93*, 182–191.

9. Castellano, G.; Lara, A.; Torrens, F. Classification of Stilbenoid Compounds by Entropy of Artificial Intelligence. *Phytochemistry* **2014**, *97*, 62–69.

10. Castellano, G.; Torrens, F. Information Entropy-Based Classification of Triterpenoids and Steroids from *Ganoderma*. *Phytochemistry* **2015**, *116*, 305–313.

11. Castellano, G.; Torrens, F. Quantitative Structure–Antioxidant Activity Models of Isoflavonoids: A Theoretical Study. *Int. J. Mol. Sci.* **2015**, *16*, 12891–12906.

12. Castellano, G.; Redondo, L.; Torrens, F. QSAR of Natural Sesquiterpene Lactones as Inhibitors of Myb-Dependent Gene Expression. *Curr. Top. Med. Chem.* **2017**, *17*, 3256–3268.

13. Torrens, F.; Castellano, G. Structure-activity relationships of cytotoxic lactones as inhibitors and mechanisms of action. *Curr. Drug Discov. Technol*, in press.

14. Torrens, F.; Redondo, L.; Castellano, G. Artemisinin: Tentative Mechanism of Action and Resistance. *Pharmaceuticals* **2017**, *10*, 20.

15. Torrens, F.; Castellano, G. A Tool for Interrogation of Macromolecular Structure. *J. Mater. Sci. Eng. B* **2014**, *4*(2), 55–63.

16. Torrens, F.; Castellano, G. Mucoadhesive Polymer Hyaluronan as Biodegradable Cationic/Zwitterionic-Drug Delivery Vehicle. *ADMET DMPK* **2014**, *2*, 235–247.

17. Torrens, F.; Castellano, G. Computational Study of Nanosized Drug Delivery from Cyclodextrins, Crown Ethers and Hyaluronan in Pharmaceutical Formulations. *Curr. Top. Med. Chem.* **2015**, *15*, 1901–1913.

18. Estrela, J. M.; Mena, S.; Obrador, E.; Benlloch, M.; Castellano, G.; Salvador, R.; Dellinger, R. W. Polyphenolic Phytochemicals in Cancer Prevention and Therapy: Bioavailability versus Bioefficacy. *J. Med. Chem.* **2017**, *60*, 9413–9436.

19. Torrens, F.; Castellano, G. From Asia to Mediterranean: Soya Bean, Spanish Legumes and Commercial *Soya Bean* Principal Component, Cluster and Meta-Analyses. *J. Nutr. Food Sci.* **2014**, *4* (5), 98.

20. Redondo-Cuevas, L.; Castellano, G.; Raikos, V. Natural Antioxidants from Herbs and Spices Improve the Oxidative Stability and Frying Performance of Vegetable Oils. *Int. J. Food Sci. Technol.* **2017**, *52*, 2422–2428.

21. Redondo-Cuevas, L.; Castellano, G.; Torrens, F.; Raikos, V. Revealing the Relationship Between Vegetable Oil Composition and Oxidative Stability: A Multifactorial Approach. *J. Food Compos. Anal.* **2018**, *66*, 221–229.

22. Torrens, F.;Castellano, G. Chemical and Biological Screening Approaches to Phytopharmaceuticals. In *Research Methods and Applications in Chemical and Biological Engineering*; Pourhashemi, A.; Deka, S.C., Haghi, A.K., Eds.; Apple Academic-CRC: Waretown, NJ, 2019, pp 3–12.

23. Torrens, F.;Castellano, G. Cultural Interbreeding in Indigenous and Scientific ethnopharmacology. In *Research Methods and Applications in Chemical and Biological Engineering*; Pourhashemi, A.; Deka, S.C., Haghi, A.K., Eds.; AppleAcademic-CRC: Waretown, NJ, 2019, pp 25–35.

24. Tan, J. W.; Israf, D. A.; Tham, C. L. Major Bioactive Compounds in Essential Oils Extracted from the Rhizomes of *Zingiber zerumbet* (L) Smith: A Mini-Review on the Anti-Allergic and Immunomodulatory Properties. *Front. Pharmacol.* **2018**, *9*, 652.

25. Leonti, M.; Casu, L. Ethnopharmacology of Love. *Front. Pharmacol.* **2018**, *9*, 567.

26. Axiotis, E.; Halabalaki, M.; Skaltsounis, L. A. An Ethnobotanical Study of Medicinal Plants in the Greek Islands of North Aegean Region. *Front. Pharmacol.* **2018**, *9*, 409.

CHAPTER 5

CHEMICAL COMPONENTS FROM *Artemisia austro-yunnanensis*, ANTI-INFLAMMATORY EFFECTS, AND LACTONES

FRANCISCO TORRENS[1,*] and GLORIA CASTELLANO[2]

[1]*Institut Universitari de Ciència Molecular, Universitat de València, Edifici d'Instituts de Paterna, P. O. Box 22085, E-46071 València, Spain*

[2]*Departamento de Ciencias Experimentales y Matemáticas, Facultad de Veterinaria y Ciencias Experimentales, Universidad Catolica de Valencia San Vicente Mártir, Guillem de Castro-94, E-46001 València, Spain*

Corresponding author. E-mail: torrens@uv.es

ABSTRACT

The essential oil from *Artemisia austro-yunnanensis* flowers was extracted by hydrodistillation, and analyzed by gas chromatography/flame ionization detector and gas chromatography-mass spectrometry, showing weak antioxidant activity by 2,2-diphenyl-1-picrylhydrazyl radical-scavenging assay. It will provide a basis for the further study on the bioactive constituents from the plant. To find bioactive constituents from *A. austro-yunnanensis* essential oil is necessary. The essential oil from flowers of the plant showed weak antioxidant activity by 2,2-diphenyl-1-picrylhydrazyl radical scavenging assay, which will provide a basis for the study on the bioactive constituents from the plant. α-Methylene-γ-lactone moiety may be the potential anti-inflammatory group in the sesquiterpene lactones.

Among the isolates from the traditionally used herbal remedy *Centaurea ragusina*, the most active compound resulted a sesquiterpene lactone, ragusinin. Though the actual protein targets remain unclear, it is confirmed that ragusinin is deactivated by glutathione (GSH) resulting in a diminished cytotoxic effect. In future studies, it will be interesting to investigate if ragusinin–GSH conjugates present a bioactivity or if they are ejected from cells via GSH pumps, for example, multidrug resistance-associated proteins MRP1 and MRP2.

5.1 INTRODUCTION

Antimalarial activity of plant metabolites was reviewed.[1] Essential oils (EOs) and their constituents were revised, targeting GABAergic system and Na^+ channels as treatment of neurological diseases.[2] Red and blue light promote the accumulation of artemisinin (ART) in *Artemisia annua* (Asteraceae).[3] *Artemisia* bioactive phenolics and their inhibitory potential versus α-amylase and α-glucosidase were informed.[4]

Genus *Artemisia* presents a variety of bioactivities. *Artemisia* is abundant in a variety of natural substances, of which sesquiterpene lactones (STLs), especially guaianolides and eudesmanolides, are the characterized ingredients, which show many obvious activities (e.g., antimalarial, anti-inflammatory, antioxidant, antitumor, antibacterial, antifungal, and antiviral). Among the active molecules, ART, from *A. annua*, is undoubtly a star STL with excellent therapeutic effect on malaria.

Earlier publications classified 31 STLs.[5,6] It was informed the tentative mechanism of action, resistance of ART derivatives (ARTDs, cf. Fig. 5.1),[7] reflections, proposed molecular mechanism of bioactivity, resistance,[8] chemical and biological screening approaches, and phytopharmaceuticals.[9] The aim of this work is to review the composition and antioxidant activity of *Artemisia austro-yunnanensis* flowers EO, highly oxidized sesquiterpenes from *A. austro-yunnanensis*, chemical components of *A. austro-yunnanensis* and their bioactivities, *A. austro-yunnanensis* 1,10-secoguaianolides and anti-inflammatory effects, and bioactivity of flavonoids and rare STLs isolated from *Centaurea ragusina*.

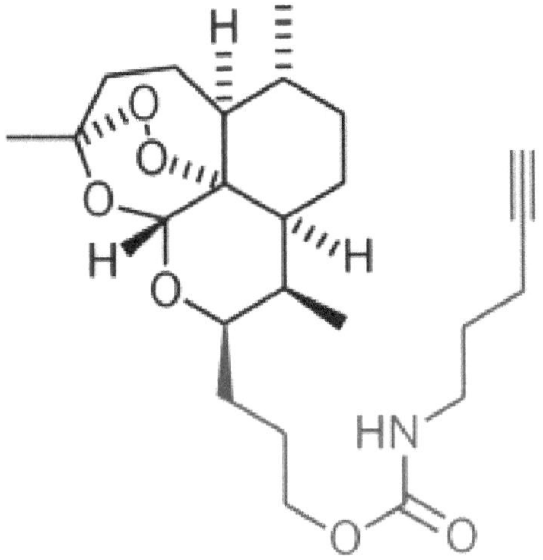

FIGURE 5.1 Molecular structure of an artemisinin derivative.

5.2 COMPOSITION AND ANTIOXIDANT ACTIVITY OF
A. austro-yunnanensis FLOWERS EO

Composition and antioxidant activity of EO from flowers of *A. austro-yunnanensis* Ling et Y. R. Ling (family: Compositae; order: Asterales) were informed.[10] The main compositions were 3,3,6-trimethyl-1,5-heptadien-4-ol (cf. Fig. 5.2a), borneol (Fig. 5.2b), (–)-caryophyllene oxide (Fig. 5.2c), agarospirol (Fig. 5.2d), humulane-1,6-dien-3-ol (Fig. 5.2e), and β-bisabolol (Fig. 5.2f), which were dominated by oxygenated monoterpenes and oxygenated sesquiterpenes. Agarospirol is a neuroleptic, and caryophyllene oxide showed significant central and peripheral analgesic and anti-inflammatory activity. Borneol specifically inhibits the nicotinic acetylcholine (ACh) receptor (nAChR)-mediated effects in a noncompetitive way. To find bioactive constituents from this plant EO is necessary.

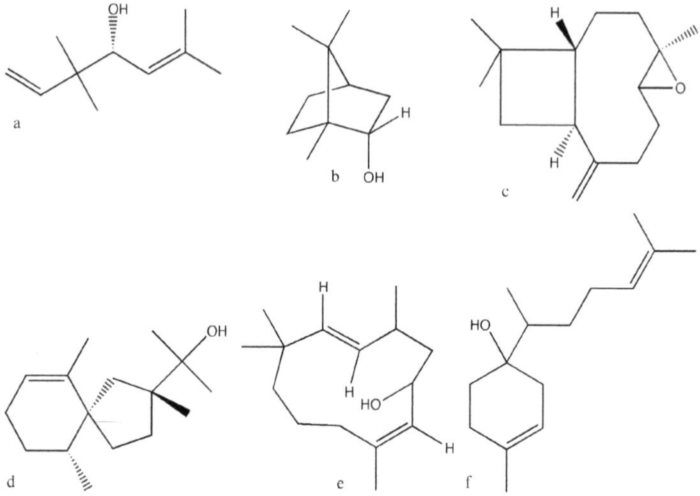

FIGURE 5.2 (a) 3,3,6-Trimethyl-1,5-heptadien-4-ol; (b) borneol; (c) caryophyllene oxide; (d) agarospirol; (e) humulane-1,6-dien-3-ol; and (f) β-bisabolol.

5.3 HIGHLY OXIDIZED SESQUITERPENES FROM *A. austro-yunnanensis*

Chemical constituents from *A. austro-yunnanensis* EO were informed.[11,12] Highly oxidized sesquiterpenes from *A. austro-yunnanensis* were reported.[13] Figure 5.3 shows the chemical structures of four guaianolides (**1–4**), one guaian (**5**), one norguaianolide (**6**), one 1,10-secoguaianolide (**7**), and one eudesmane (**8**) sesquiterpenes.

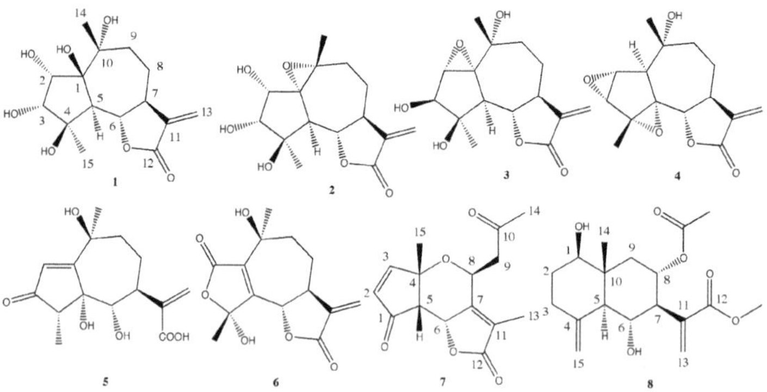

FIGURE 5.3 Chemical structures of seven guaiane 1–7 and one eudesmane 8 sesquiterpenes.

5.4 CHEMICAL COMPONENTS OF *A. austro-yunnanensis* AND THEIR BIOACTIVITIES

Chemical components and their bioactivities of *A. austro-yunnanensis* EO were informed.[14] Figure 5.4 shows five lignans (**9–13**) and five phenolics (**14–18**) isolated from *A. austro-yunnanensis*: 7-(3-ethoxy-5-methoxy-phenyl)propane-7,8,9-triol (**9**), secoisolariciresinol (**10**), (+)-pinoresinol (**11**), syringaresinol (**12**), eugenyl-*O*-β-D-glucopyranoside (**13**), benzyl-*O*-β-D-glucopyranoside (**14**), *p*-hydroxybenzaldehyde (**15**), vanillin (**16**), syringaldehyde (**17**), and 3-hydroxy-1-(4-hydroxy-3,5-dimethoxyphenyl)-1-propanone (**18**).

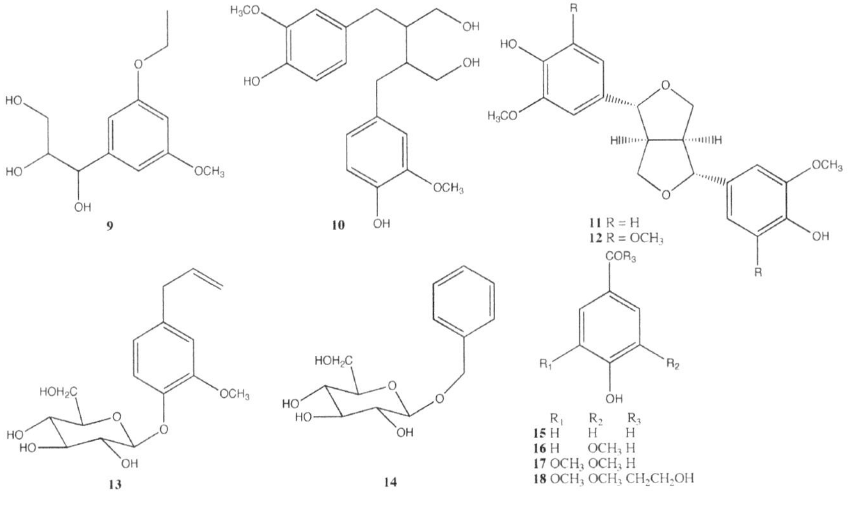

FIGURE 5.4 Molecular structures of compounds isolated from *Artemisia austro-yunnanensis*.

5.5 *A. austro-yunnanensis* 1,10-SECOGUAIANOLIDES AND ANTI-INFLAMMATORY EFFECTS

1,10-Secoguaianolides from *A. austro-yunnanensis* EO and their anti-inflammatory effects were informed.[15] Figure 5.5 shows seven 1,10-secoguaianolides **19–25**.

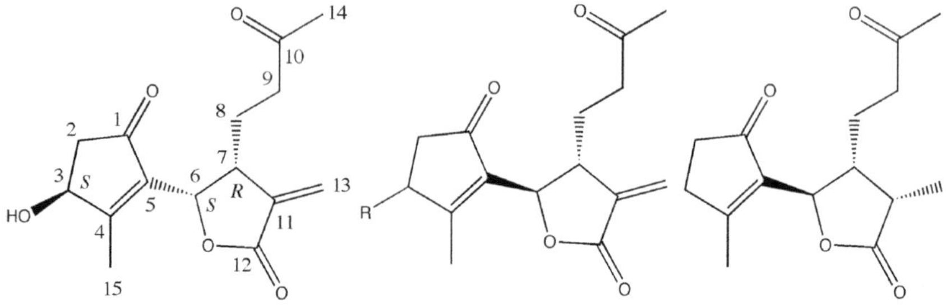

19

20 R = α-OH
21 R = β-OH
22 R = α-OMe
23 R = β-OMe
24 R = β-OEt

25

FIGURE 5.5 Chemical structures of 1,10-secoguaianolides 19–25.

5.6 BIOACTIVITY OF FLAVONOIDS AND RARE STL ISOLATED FROM *C. ragusina*

Phytochemical and biopotential of in vivo and in vitro *C. ragusina* L., and detection of deoxyribonucleic/ribonucleic acid (DNA/RNA) active compounds in extracts via thermal denaturation and circular dichroism were informed.[16] The bioactivity of flavonoids and rare STLs from *C. ragusina* leaves were reported.[17] Six isolates (cf. Fig. 5.6) were identified as chrysin (**26**), oroxylin A (**27**), hispidulin (**28**), deacylcynaropicrin (**29**), (3aR,4S,6aR,8S,9aR,9bR)-[dodecahydro-8-dihydroxy-3,6,9-tris(methylene)-2-oxo-2(3*H*)-azuleno[4,5-b]furanyl]-3-methylbutanoate (ragusinin, **30**), and hemistepsin A (**31**). The flavonoids (**26–28**) can be classified as flavones, whereby different substitution patterns with methoxy or hydroxy groups can be observed on C-6 and C-4′. Compounds **29–31** are STLs belonging to the subtype of guaianolides.

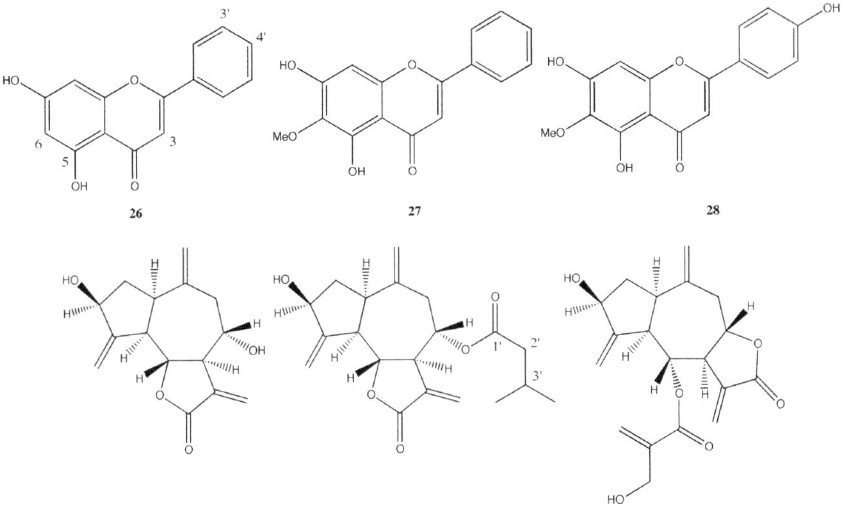

FIGURE 5.6 Chemical structures of isolated *Centaurea ragusina* L. leaf constituents (26–31).

5.7 DISCUSSION

Artemisia austro-yunnanensis flowers EO was extracted by hydrodistillation, and analyzed by gas chromatography/flame ionization detector (GC-FID) and GC-mass spectrometry (MS), showing weak antioxidant activity by 2,2-diphenyl-1-picrylhydrazyl (DPPH) radical-scavenging assay. It will provide a basis for the further study on the bioactive constituents from this plant.

All isolated sesquiterpenes and a series of extracts of *A. austro-yunnanensis* were tested their potential anti-inflammatory activities on the model of lipopolysaccharide (LPS)-induced nitric oxide (NO) production in RAW 264.7 cell line, of which compounds **2–4** produced significant inhibition of NO production with concentration for 50% inhibition (IC_{50}) values in 4.20–10.67 µM. α-Methylene-γ-lactone (ML) moiety was found the important functional group in guaiane and eudesmane susquiterpenes with antitumor and antibacterial activities. The bioassay results showed compounds **2–4** with ML moiety, except **1** and **6** with no activities, produce significant inhibition of NO production, speculating ML moiety may be the potential anti-inflammatory group in STLs.

Lignan **10** exhibited better inhibiting neutrophil elastase effect at 100 μM than that of other compounds. As a special semi-herbaceous shrub in genus *Artemisia*, the researches of chemical constituents and activities of *A. austro-yunnanensis* were limited. The research can provide foundation for development of the plant and investigation of its phytotaxonomy.

Six 1,10-secoguaianolides **19–22**, **24**, and **25** isolated from *A. austro-yunnanensis* produced obvious anti-inflammatory effects via decreasing the release of NO, tumor necrosis factor-α (TNF-α), interleukin (IL)-1β, IL-6, and prostaglandin (PG) E2 (PGE2), and down-regulating the expression of proteins inducible NO synthase (iNOS) and cyclooxygenase-2 (COX-2). They regulated the nuclear factor-κB (NF-κB)-dependent transcriptional activity via decreasing the phosphorylation of NF-κB.

Among the isolates (**26–31**) from the traditionally used herbal remedy *C. ragusina* leaves, an interesting pharmacological profile was discovered for the rare guaianolides ragusinin (**30**) and hemistepsin A (**31**). Compound **30** was described as a constituent from the aerial parts of the Australian *Helipterum maryonii* S. Moore and is characterized by an isovalerate residue in position C-9a; substance **31** was described as a constituent from leaves and flowers of *Hemisteptia lyrata* Bunge. Glutathione (GSH) is the ubiquitous peptide that traps reactive compounds and other xenobiotics to prevent damage to vital proteins and nucleic acids. The variation of GSH amount in the cell induced by a specific precursor in GSH synthesis, *N*-acetylcysteine (NAC), or a specific inhibitor of GSH synthesis, L-buthionine sulphoximine (BSO), presented an impact on Henrietta Lacks (HeLa) survival upon treatment with **30**. Similar results were obtained in human colon tumor cells upon BSO and pseudoguaianolide STL helenalin treatment. Moreover, GSH probably does not play a role as antioxidant but rather as molecule which form conjugates with **30** increasing in that way the survival of HeLa cells upon treatment with it.

5.8 FINAL REMARKS

From the previous results, the following final remarks can be drawn.

1. To find bioactive constituents from *A. austro-yunnanensis* essential oil is necessary. The essential oil from flowers of this plant showed weak antioxidant activity by DPPH radical scavenging

assay, which will provide a basis for the further study on the bioactive constituents from this plant.

2. ML moiety may be the potential anti-inflammatory group in the sesquiterpene lactones.

3. Among the isolates (**26–31**) from the traditionally used herbal remedy *C. ragusina*, the most active compound resulted a sesquiterpene lactone, ragusinin (**30**). Though the actual protein targets remain unclear, it is confirmed that **30** is deactivated by GSH resulting in a diminished cytotoxic effect.

4. In the future studies, it will be interesting to investigate whether ragusinin–GSH conjugates present a bioactivity or whether they are ejected from cells via glutathione pumps, for example, multidrug resistance-associated proteins MRP1 and MRP2.

ACKNOWLEDGMENTS

The authors thank support from Generalitat Valenciana (Project No. PROMETEO/2016/094) and Universidad Católica de Valencia *San Vicente Mártir* (Project Nos. UCV.PRO.17-18.AIV.03 and 2019-217-001).

KEYWORDS

- 1,10-secoguaianolide
- anti-inflammation
- NF-κB regulation
- iNOS inhibition
- lignan
- phenolic
- bioactivity

REFERENCES

1. Pan, W. H.; Xu, X. Y.; Shi, N.; Tsang, S. W.; Zhang, H. J. Antimalarial Activity of Plant Metabolites. *Int. J. Mol. Sci.* **2018**, *19*, 1382.

2. Wang, Z. J.; Heinbockel, T. Essential Oils and Their Constituents Targeting the GABAergic System and Sodium Channels as Treatment of Neurological Diseases. *Molecules* **2018**, *23*, 1061.

3. Zhang, D.; Sun, W.; Shi, Y.; Wu, L.; Zhang, T.; Xiang, L. Red and Blue Light Promote the Accumulation of Artemisinin in *Artemisia annua* L. *Molecules* **2018**, *23*, 1329.

4. Olennikov, D. N.; Chirikova, N. K.; Kashchenko, N. I.; Nikolaev, V. M.; Kim, S. W.; Vennos, C. Bioactive Phenolics of the Genus *Artemisia* (Asteraceae): HPLC-DAD-ESI-TQ-MS/MS Profile of the Siberian Species and Their Inhibitory Potential Against α-Amylase and α-Glucosidase. *Front. Pharmacol.* **2018**, *9*, 756.

5. Castellano, G.; Redondo, L.; Torrens, F. QSAR of Natural Sesquiterpene Lactones as Inhibitors of Myb-Dependent Gene Expression. *Curr. Top. Med. Chem.* **2017**, *17*, 3256–3268.

6. Torrens, F.; Castellano, G. Structure-activity relationships of cytotoxic lactones as inhibitors and mechanisms of action. *Curr. Drug Discov. Technol*, in press.

7. Torrens, F.; Redondo, L.; Castellano, G. Artemisinin: Tentative Mechanism of Action and Resistance. *Pharmaceuticals* **2017**, *10*, 20.

8. Torrens, F.; Redondo, L.; Castellano, G. Reflections on Artemisinin, Proposed Molecular Mechanism of Bioactivity and Resistance. In *Applied Physical Chemistry with Multidisciplinary Approaches*; Haghi, A. K., Balköse, D., Thomas, S., Eds.; Apple Academic–CRC: Waretown, NJ, 2018; pp 189–215.

9. Torrens, F.;Castellano, G. Chemical and Biological Screening Approaches to Phytopharmaceuticals. In *Research Methods and Applications in Chemical and Biological Engineering*; Pourhashemi, A.; Deka, S.C., Haghi, A.K., Eds.; Apple Academic-CRC: Waretown, NJ, 2019, pp 3–12.

10. Zhao, C. X.; Zhang, M.; He, J.; Ding, Y. F.; Li, B. C. Chemical Composition and Antioxidant Activity of the Essential Oil from the Flowers of *Artemisia austro-yunnanensis*. *J. Chem. Pharm. Res.* **2014**, *6*, 1583–1587.

11. Chi, J.; Li, B. C.; Zhang, M. Chemical Constituents from *Artemisia austro-yunnanensis*. *J. Kunming Univ. Sci. Tech. (Nat. Sci. Ed.)* **2015**, *40*, 93–96.

12. Chi, J.; Dai, W. F.; Li, B. C.; Qin, Y.; Jiao, S.; Zhang, M. Chemical Constituents from *Artemisia austro-yunnanensis* (II). *J. Kunming Univ. Sci. Tech. (Nat. Sci. Ed.)* **2017**, *42*, 2–17.

13. Chi, J.; Li, B. C.; Dai, W. F.; Liu, L.; Zhang, M. Highly Oxidized Sesquiterpenes from *Artemisia austro-yunnanensis*. *Fitoterapia* **2016**, *115*, 182–188.

14. Chi, J.; Li, B. C.; Yang, B. T.; Zhang, M. Chemical Components and Their Bioactivities of *Artemisia austro-yunnanensis*. *J. Chem. Soc. Pak.* **2016**, *38*, 533–537.

15. Liu, L.; Dai, W.; Xiang, C.; Chi, J.; Zhang, M. 1,10-Secoguaianolides from *Artemisia austro-yunnanensis* and Their Anti-Inflammatory Effects. *Molecules* **2018**, *23*, 1639.

16. Vujcic, V.; Radic Brkanac, S.; Radojcic Redovnikovic, I.; Ivankovic, S.; Stojkovic, R.; Zilic, I.; Radic Stojkovic, M. Phytochemical and Bioactive Potential of In Vivo and In Vitro Grown Plants of *Centaurea ragusina* L.—Detection of DNA/RNA Active

Compounds in Plant Extracts via Thermal Denaturation and Circular Dichroism. *Phytochem. Anal.* **2017**, *28*, 584–592.

17. Grienke, U.; Radic Brkanac, S.; Vujcic, V.; Urban, E.; Ivankovic, S.; Stojkovic, R.; Rollinger, J. M.; Kralj, J.; Brozovic, A.; Radic Stojkovic, M. Biological Activity of Flavonoids and Rare Sesquiterpene Lactones Isolated from *Centaurea ragusina* L. *Front. Pharmacol.* **2018**, *9*, 972.

CHAPTER 6

SPECTROSCOPIC AND LATTICE PROPERTIES OF LANTHANIDE COMPLEX OXIDES

DIMITAR N. PETROV*

Department of Physical Chemistry, Faculty of Chemistry, Paisii Hilendarski University, 24 Tzar Asen Str., 4000 Plovdiv, Bulgaria

E-mail: petrov_d_n@abv.bg

ABSTRACT

The present chapter deals with stoichiometric lanthanide and doped rare-earth complex oxides. It contains a survey of recent studies on certain basic properties and applications of $4f^N$ electron systems. The following topics are included: lattice energies, dielectric properties, magnetic susceptibilities, and magnetic exchange interactions, optical spectra, nephelauxetic effect, and derivation of radial expectation values for Ln^{3+} ions, two-center overlap with participation of 4f electron wave functions, crystalline materials of nanosize particles.

6.1 INTRODUCTION

In order to comply with the meaning of the word "complex" in the title, the objects of consideration will be restricted mainly to double and ternary oxide systems containing Ln^{3+} ions as stoichiometric constituents and, rarely, as dopants (activators). Examples of such oxide systems with lanthanides comprise $LnEO_3$, $Ln_3E_5O_{12}$, with E = Al, Ga or Fe, $LnPO_4$, $LnVO_4$, $LiLnP_4O_{12}$, $LnAl_3(BO_3)_4$, etc. Lanthanides (14 elements) are part of the rare earths: the latter include also scandium ($_{21}Sc$), yttrium ($_{39}Y$), and lanthanum ($_{57}$La). Thus, the group of rare earths consists of 17 elements which have got common discovery, separation, and similar properties. Often lanthanides have been referred as rare earths in the literature.

The series of rare earth garnets $R_3E_5O_{12}$ where E is Fe or Ga, as one of the most thoroughly studied group of ferrites, have been assessed by thermodynamic treatment of their static and vibrational properties.[1] It has been also found that their mixing (or fractionations) at various pressures and temperatures are determined primarily by the energy characteristics of $R_3E_5O_{12}$ end members.

Rare earth aluminates $RAlO_3$ and gallates $RGaO_3$ have been studied for their valuable physical and chemical properties and numerous applications.[2] The emphasis of the same chapter is on the structure analysis and the variations of the perovskite-type structures with temperature, pressure, and stoichiometry. The structure analysis has included: bond-length distortions, angles between octahedra, tolerance in various coordination numbers, ion–ion distances, etc.

The following overview is oriented to a greater extent toward recent achievements in spectroscopic parameterization, lattice energies, bonding effects, and applications of complex oxides rather than to contemporary structural elucidations. There exist at least three arguments based on the properties of the mentioned systems that justify the continuous research efforts in the field: (1) stability—thermal, chemical, and mechanical; (2) variety of appropriate optical, magnetic, and dielectric characteristics; and (3) technological applications.

6.2 CERTAIN CRYSTAL STRUCTURES

6.2.1 YTTRIUM ALUMINUM GARNET, $Y_3Al_5O_{12}$

YAG is a hitech material and a brief description of its structure and properties is outlined below. It belongs to the garnets with formula $A_3B_5O_{12}$, where A = Y, Ln or U, B = Al, Cr, Fe, etc. $A_3B_5O_{12}$ originated from garnets which are orthosilicates with general formula $A^{II}_3B^{III}_2(SiO_4)_3$, where A^{II} = Ca^{2+}, Mg^{2+} or Fe^{2+}, and B^{III} = Al^{3+}, Cr^{3+}, or Fe^{3+}. By substituting A^{II} with Y^{3+} and Si^{4+}—with Al^{3+}, $Y_3Al_5O_{12}$ is derived, thus preserving the overall net charge. The molar ratio $Y_2O_3:Al_2O_3 = 3:5$, that is, $Y_3Al_5O_{12}$ composition is half the full stoichiometry based on the same oxides.

Yttrium aluminum garnet crystallizes in a complex, cubic structure, in which one unit cell contains eight formula units (or eight moles $Y_3Al_5O_{12}$). Thus, cations Al^{3+} occupy 16-fold octahedral sites, AlO_6 (designated as 16a) and 24-fold tetrahedral sites, AlO_4 (or 24d). Y^{3+} cations occupy 24-fold dodecahedral sites, YO_{12} (or 24c). All these polyhedra, formed by oxygen ions, O^{2-} are distorted and rotated in respect to certain axes: octahedra, by angle $\pm\beta \approx 28°$, tetrahedra, by angle $\pm\beta \approx 15.6°$.

Each dodecahedron has common O–O edges with two tetrahedra, four octahedral, and four other dodecahedra. Elements of the crystal structure of $Y_3Al_5O_{12}$ are presented in Fig. 6.1 (Table 6.1).

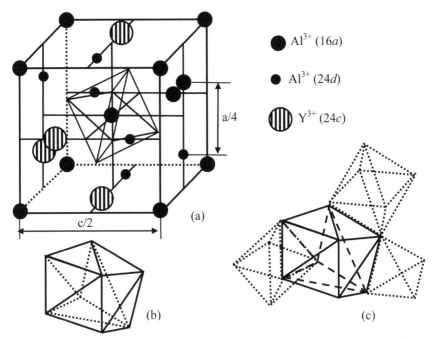

FIGURE 6.1   Structure of yttrium aluminum garnet, $Y_3Al_5O_{12}$: (a) cationic positions, (b) trigondodecahedron YO_{12}, and (c) matching dodecahedron YO_{12}, octahedron AlO_6, and tetrahedron AlO_4.

TABLE 6.1   Distances (in pm) in Yttrium Aluminum Garnet, $Y_3Al_5O_{12}$.

4(Al–O)	6(Al–O)	8(Y–O)	4,8(O–O)	6,8(O–O)	8,8(O–O)
176.1	193.7	243.2	269.6	265.8	283.7

Lengths of the common edges O–O: 4,8—tetrahedron and dodecahedron, 6,8—octahedron and dodecahedron, 8,8—two dodecahedra.

$Y_3Al_5O_{12}$ *basic data:*

Molar mass, $M = 593.59 \times 10^{-3}$ kg mol⁻¹; density, $\rho = 4.55 \times 10^3$ kg m⁻³;melting point, $T_m = 2203 \pm 20$ K; unit cell constant, $a = 1.2008$ nm;space group: $O_h^{10} - Ia3d$; hardness (Mohs): 8.5; relative dielectric permittivity: $\varepsilon_0 = 11.7$, $\varepsilon_\infty = 3.5$; refractive index ($\lambda = 1.064$ μm): 1.81523.

6.2.2 GADOLINIUM ALUMINATE, GdAlO₃ SPACE GROUP: D¹⁶₂ₕ – Pbnm

Gadolinium aluminate is another hitech material that pertains to the lanthanide monoaluminates LnAlO₃, for Ln = Sm – Lu, with orthorhombic, perovskite-type structure GdFeO₃; its orthorhombic structure does not undergo phase transition toward rhombohedral one in the temperature range 14–1170 K.[2]

Many compounds ABX₃ have distorted perovskite structure into an orthorhombic one and Fig. 6.2 contains examples of orthorhombic and rhombohedral types of these distortions.

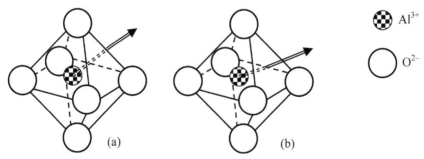

FIGURE 6.2 Orthorhombic (a) and rhombohedral (b) distortions in octahedra AlO₆ of LnAlO₃.

The orthorhombic structure of GdAlO₃ is presented in Fig. 6.3. Gd³⁺ ions have CN = 12, whereas Al³⁺ ions, CN = 6; the polyhedron GdO₁₂ is a cube octahedron, while AlO_6 is octahedron. The Gd—O bond is predominantly ionic but the Al—O bond is basically covalent.

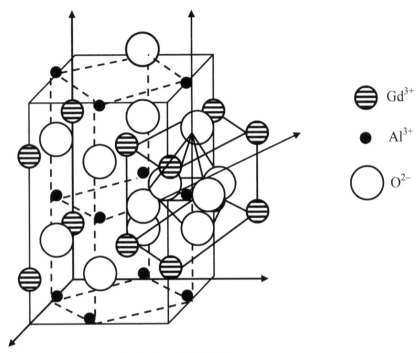

FIGURE 6.3 Unit cell of orthorhombic $GdAlO_3$.

$LiLnP_4O_{12}$ have monoclinic crystal structure, space group C2/c. The unit cell contains four molecules ($z = 4$) and the site symmetry of the Ln^{3+} ions is 2 (or C_2). Certain distances in $LiNdP_4O_{12}$ (in pm): Nd–Nd, 562; Nd–O, 2 × 256, 2 × 242, 2 × 249, 2 × 238.[4]

6.2.2 MAGNETIC PROPERTIES

The Ln^{3+} ions either in a free state or in an insulating crystal are paramagnetic at room temperature (RT) because of the unpaired 4f electrons. The order of energy contributions to the Hamiltonian is the following:

$$E_E > E_{SO} > E_{CF} \approx kT. \qquad (6.1)$$

The paramagnetic susceptibility χ and the effective magnetic moments μ_{eff} of N noninteracting Ln^{3+} ions in their ground levels, at RT and below, may be determined by the Hund's equation:

$$\chi = Ng_J^2\mu_B^2 J(J+1)/3 \, kT, \quad \mu_{eff} = g \, [J(J+1)]^{1/2}. \qquad (6.2)$$

The IUPAC recommendations for magnetic properties require presentation of all quantities, dimensions and conversion factors in SI.[27] The plot of inverse molar magnetic susceptibility versus T contains the Curie constant C in the slope and the Weiss constant θ in the intercept:

$$\chi_m^{-1} = (1/C) \, T + (1/C) \, \theta, \quad C = (\mu_0 N_A \mu_B^2 \mu_{eff}^2 / 3k_B). \qquad (6.3)$$

The effective Bohr magneton number of Ln^{3+} is dimensionless as the dimension of μ_B is $JT^{-1} = A \, m^2$:

$$\mu_{eff} = (3k_B C/\mu_0 N_A \mu_B^2)^{1/2}. \qquad (6.4)$$

Five orthorhombic lanthanide monoaluminates (OLA) $LnAlO_3$, Ln = Nd, Sm, Eu, Gd, Dy, have been obtained by low-temperature nanosynthesis with the aid of malic acid. All samples have been characterized by X-ray diffraction (XRD), scanning electron microscopy (SEM), transmission electron microscopy (TEM), photoelectron spectroscopy, infrared (IR) spectroscopy, electron spin resonance ($GdAlO_3$). Magnetic studies have been performed on PPMS-9 in the temperature range 2–300 K.[28–32] The following magnetic quantities and properties have been determined (Fig. 6.5):

1. the effective magnetic moment for each Ln^{3+} ion from the experimental dependence of inverse molar magnetic susceptibility on T applying eqs (6.3) and (6.4);
2. the constants of exchange interactions between $GdAlO_3$ and $DyAlO_3$;
3. the magnetization dependence $M = f(H)$ for $GdAlO_3$ and $DyAlO_3$;
4. the Neél temperature T_N, that is, the temperature of magnetic phase transition "paramagnetism $\leftrightarrow$ antiferromagnetism" when that transition takes place at temperatures $T \geq 2$ K, as it is in the case for $GdAlO_3$ and $DyAlO_3$;

5. the theoretical curve for the molar magnetic susceptibility versus T for $EuAlO_3$.

An example of nanocrystalline $DyAlO_3$ is given in Figure 6.6.

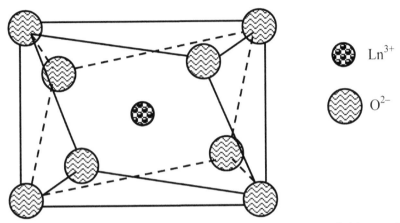

FIGURE 6.4   Polyhedron LnO_8 in $LiLnP_4O_{12}$. Note that the symmetry axis 2 is normal to the symmetry plane m, that is, $2/m$.
Source: Adapted with permission from Ref. [3] with revisions.

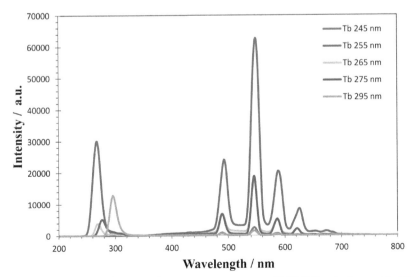

FIGURE 6.5   **(See color insert.)** Fluorescence spectra of $LiTbP_4O_{12}$ single crystal at RT after excitations with LEDs between 255 and 295 nm.

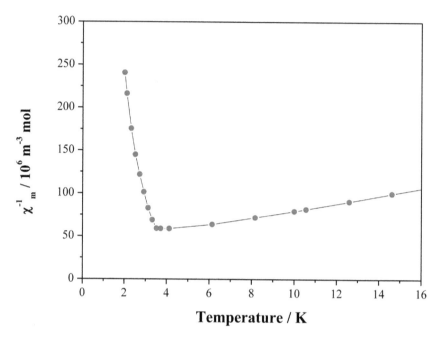

FIGURE 6.6   **(See color insert.)** Variation of the inverse molar DC magnetic susceptibility $\chi_m^{-1}/10^6$ mol m^{-3} of Dy^{3+}ions in nanocrystalline DyAlO$_3$ at $T = 2$–15 K and $H = 1000$ Oe.

The model of Anderson for magnetic exchange in insulators[33] has been applied to OLA.[34] The arguments supporting such a treatment are based on the following experimental facts: the Ln^{3+} – Ln^{3+} distances are shorter than the Ln^{3+} – O^{2-} – Ln^{3+} distances, there exist spin-dependent low-lying electron states in Ln^{3+} ions, LnAlO$_3$ are not diluted but stoichiometric compounds with molar ratio 1:1 of the corresponding sesquioxides. The constants J_{ex} of magnetic exchange interactions have been evaluated and used further in the calculation of the Neél temperatures in nine OLA within the molecular field approximation; a comparison with the experimental values is presented in Figure 6.7.

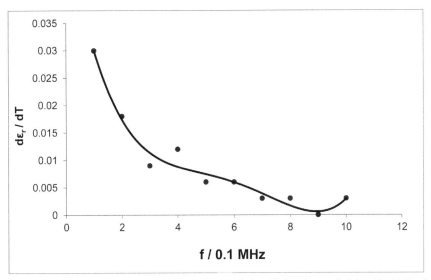

FIGURE 6.7   Temperature coefficient of the relative dielectricpermittivity ($\Delta\varepsilon_r/\Delta T$) of nanocrystalline GdAlO$_3$ in the range 298–323 K versus frequency f.

6.2.3   DIELECTRIC STUDIES

The Maxwell equations relate magnetic flux density B, current density J, the strength of the electric field E with the permeability μ and dielectric permittivity ε:

$$(1/\mu) \times B = J + \varepsilon\, (\partial E/\partial t) \times E = (\partial B/\partial t); \qquad (6.5)$$

The current density J is related to the strength of the electric field E by the conductivity σ:

$$J = \sigma\, E. \qquad (6.6).$$

The corresponding scalar form of J includes also the angular velocity or frequency) ω:

$$J - i\,\omega\,\varepsilon\,E = \sigma\,E - i\,\omega\,\varepsilon\,E = -i\,\omega\,[\varepsilon + I\,(\sigma/\omega)]\,E. \qquad (6.7).$$

The term in square brackets is referred to as complex dielectric permittivity.

$$E = \varepsilon_1 \varepsilon_0, \; \sigma/\omega = \varepsilon_{II} \varepsilon_0. \quad\quad (6.8).$$

The loss tangent is defined as follows:

$$\tan \delta = \varepsilon_{II}/\varepsilon_I. \quad\quad (6.9).$$

Nanocrystalline $GdAlO_3$ with particle size of 100 nm has been studied by radio-frequency dielectric spectroscopy.[35] Capacitance, impedance, loss tangent, phase angle, and relative dielectric permittivity have been measured in the range 20 Hz to 1 MHz at $T = 298–473$ K. The polarizability volume of $GdAlO_3$ has been also determined at 1 MHz. Different values have been obtained for most of the dielectric quantities compared to those in another article on nanosized $GdAlO_3$.[36] Examples of the variations of the temperature coefficients of the studied dielectric quantities are given in Figures 6.8 and 6.9.

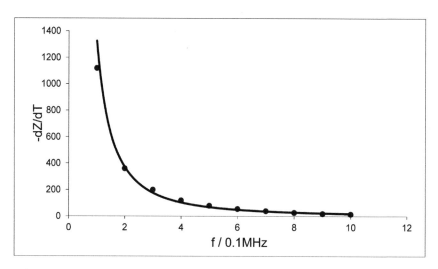

FIGURE 6.8 Temperature coefficient of the impedance $(-\Delta Z/\Delta T)$ of nanocrystalline $GdAlO_3$ in the range 298–323 K versus frequency f.

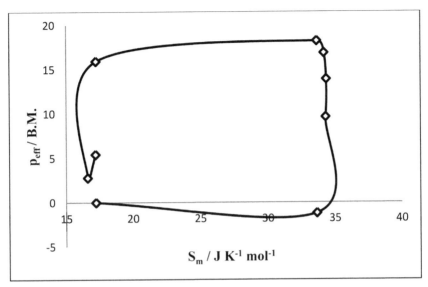

FIGURE 6.9 Loop in the series of $Ln_3Fe_5O_{12}$: experimental magnetic moments p_{eff}/B.M. versus calculated magnetic entropies S_m/J K^{-1} mol^{-1}; points clockwise, from left to right: Ln = Sm, Eu, Gd, Tb, Dy, Ho, Er, Tm, Yb.
Source: Adapted with permission from Ref. [44].

Dielectric thin films of amorphous $LaAlO_3$ have been prepared at annealing temperatures (higher than 773 K) from aqueous solutions of the corresponding nitrates.[37] The samples have been characterized by dynamic light scattering, Raman spectroscopy, leakage-current densities, average mobilities, and turn-over voltages; ε_r varies from 11.0 to 11.5.

Ceramics with composition $[R_{mix}PO_4]$ has been fabricated from natural mixture of R_2O_3 intended to low-cost, microwave-substrate applications.[38] Sintering of the rare-earth oxide mixture in the monazite sand after removal of the radioactive Th with $NH_4H_2PO_4$ at $T = 1523$ K has resulted in ceramics with $\varepsilon_r = 9.6$ at 13.5 GHz.

6.3 OPTICAL AND MAGNETIC 4f-"FINGERPRINTS"

The 4f-"fingerprints" are obviously the sharp, of low intensity bands in absorption and emission spectra of Ln^{3+} ions, on one side, and the values of magnetic moments of the same ions almost unaffected by the surrounding ions, on the other.

6.3.1 OPTICAL SPECTRA

The ground-state electron configuration of Ln^{3+} ions is $[_{54}Xe]$ $4f^N$ where N varies from 1 (Ce^{3+}) to 14 (Lu^{3+}). Often Pm^{3+} with atomic number $Z = 61$ is omitted from discussion in the lanthanide series since it has been synthesized artificially by fusion of uranium and neutron bombardment of neodymium[5]; Lu^{3+} ($Z = 61$) is often excluded when lanthanides with open-subshell $4f^N$ only have to be considered. The optical 4f-"fingerprints" are related to the 4f-energy levels usually designated according to the RS coupling scheme as $^{2S+1}L_J$ where the angular momenta S, L, and J are in units of $\hbar$.[6]

In first approximation, (1) the positions of the spectral bands of Ln^{3+} ions in a crystal are considered close to those of the free ions Ln IV, that is, very slightly affected by the surrounding ions, (2) the energy of a 4f $\leftrightarrow$ 4f transition is determined mainly by the energy of interelectron (E_E) and spin–orbit (E_{SO}) interactions:

$$E = E_E + E_{SO} = \sum_k f^k F^k (4f, 4f) + A_{SO}\zeta_{4f} \tag{6.10}$$

where $k = 2, 4, 6, f^k$, and A_{SO} represent angular parts of the interactions; the term with $k = 0$ does not contribute to the energy of the transitions between two levels. The Slater–Condon radial integrals F^k present electrostatic interactions between two electrons i and j through the radial parts $R_{4f}(r)$ of one-electron 4f wave functions[7]; F^k are treated as semiempirical parameters or calculated.

$$F^k = 2e^2 \int_0^\infty \int_0^\infty \frac{r_<^k}{r_>^{k+1}} R_{4f^2}(r_i) R_{4f^2}(r_j) dr_i dr_j \tag{6.11}$$

$$F_k = F^k / D_k \tag{6.12}$$

The F^k and F_k parameters are interrelated by means of the coefficients D_k according to Condon and Shortley[8]: $D_2 = 225$, $D_4 = 1089$, $D_6 = 7361.64$; $r_<$ and $r_>$ designate the lesser and the greater of the distances r_i and r_j, respectively.

The identification of the 4f-energy levels of Ln^{3+} ions is a major task of the analysis of their electronic spectra. It is necessary for the application of these ions in various materials.

The energy levels of Ln^{3+} ions can be derived semiempirically or theoretically.[9] In the first method, the following consecutive steps should be performed: (1) fitting the levels obtained from optical spectra to find a set of spectroscopic parameters F_2, F_4, F_6, and ζ_{4f}; (2) diagonalization of the matrices of electrostatic and spin-orbit interactions; and (3) construction of the eigen states with eigenvectors and eigen values in intermediate coupling for each level $^{2S+1}L_J$. The procedure is simple for $Pr^{3+} = [_{54}Xe]\,4f^2$ and $Tm^{3+} = [_{54}Xe]\,4f^{12}$ as far as each of these ions exhibits only 13 4f-energy levels and the order of the matrices to be solved does not exceed three. For the middle of the lanthanide series (Eu^{3+}, Gd^{3+}, and Tb^{3+}), the maximum order of the matrices of the electrostatic interactions for diagonalization is 9 whereas that of spin–orbit interactions is 46–50.[10]

The maximum number (N_T) of terms and levels within the $4f^N$ electron configuration is presented in Table 6.2.[9-11] Excluding Lu in it, with closed configuration $4f^{14}$, the number of terms and levels is symmetrical in respect to the middle of the lanthanide series at $Gd = [_{54}Xe]\,4f^7$ since the electron configurations $4f^N$ and $4f^{14-N}$ are complementary toward $N = 14$ via the number of electrons and "holes."

TABLE 6.2 Total Number (N_T) of Possible Terms (^{2S+1}L) and Levels ($^{2S+1}L_J$) for $4f^N$ Electrons with Inclusion of Electrostatic and Spin–Orbit Interactions and Values of Parameters Determining the Magnetic Susceptibility of Free Ln^{3+}Ions in Their Ground Levels.

$4f^N$, $N =$	Ln^{3+}	N_T of ^{2S+1}L	N_T of $^{2S+1}L_J$	S	L	J	g	μ_{eff}
1	Ce	1	2	1/2	3	5/2	6/7	2.54
2	Pr	7	13	1	5	4	4/5	3.58
3	Nd	17	41	3/2	6	9/2	8/11	3.62
4	Pm	47	107	2	6	4	3/5	2.68
5	Sm	73	198	5/2	5	5/2	2/7	0.84
6	Eu	119	295	3	3	0	–	0
7	Gd	119	327	7/2	0	7/2	2	7.94
8	Tb	119	295	3	3	6	3/2	9.72
9	Dy	73	198	5/2	5	15/2	4/3	10.65
10	Ho	47	107	2	6	8	5/4	10.61
11	Er	17	41	3/2	6	15/2	6/5	9.58
12	Tm	7	13	1	5	6	7/6	7.56
13	Yb	1	2	1/2	3	7/2	8/7	4.54
14	Lu	1	1	0	0	0	–	0

The lowest levels above the ground level $^{7}F_0$ of Eu^{3+} have $J = 1, 2$, and $g = 3/2, 3/2$, respectively.

In intermediate coupling approximation, the wave function of a particular eigen state (designated in square brackets) is expressed as a linear combination of individual states defined within the LS coupling scheme which contribute with coefficients C_i (eigenvectors) obeying the condition for normalization:

$$\left| [SLJ] \right\rangle = \sum_i C_i \left| (SL)J \right\rangle, \sum_i C_i^2 = 1 \qquad (6.13)$$

The increase of the number of parameters from four (F_2, F_4, F_6, and ζ_{4f}) with inclusion of two-particle (α, β, and γ) and three-particle (T^2, T^3, T^4, T^6, T^7, and T^8) configuration interactions of two-particle pseudomagnetic operators (P^2, P^4, and P^6), of spin-other-orbit, and of spin–spin effects (Marvin integrals: M^0, M^2, M^4) reduces the standard deviation between experiment and theory by 20–30% and is physically justified. Nevertheless, the increase of the number of parameters faces the following opposite considerations:

i.   these additional parameters change insignificantly in the lanthanide series;

ii.   represent sums of terms with different signs and compatible magnitudes, thus resulting in a delicate balance;

iii. the various parameters are rather sensitive toward the inclusion or the absence of certain levels of Ln^{3+} ions;

iv.   the number of parameters depends on the number of experimentally determined levels $^{2S+1}L_J$.

The parameters and the energy levels E_i from experiment and theory are subjected to an iterative procedure minimizing the root–mean–square (RMS) deviation:

$$RMSD = \left[\sum_i \left(E_{i,exp} - E_{i,calc} \right)^2 / (n-m) \right]^{1/2} = min, \qquad (6.14)$$

where n is the number of levels, m is the number of varied parameters.

One fundamental problem of the optical spectroscopy of Ln^{3+} ions is the extent of deviation from pure LS-coupling scheme. This effect has been verified quantitatively for the free ions Pr IV ($4f^2$) and Tm IV ($4f^{12}$) by evaluation of the squared matrix elements of the tensor operator $\left| U^{(k)} \right|^2$,

k = 2, 4, 6, in both coupling schemes: intermediate and LS.[12] It has been found that among all 4f–4f transitions only those lower than 7000 cm^{-1} for Pr IV or below 15,000 cm^{-1} for T_m IV have close values of $\left| U^{(k)} \right|^2$ in intermediate and LS couplings.

Energy-level diagrams have been deduced theoretically for Ln^{2+}, Ln^{3+}, and Ln^{4+} ions with the aid of Hartree–Fock calculated electrostatic and spin–orbit interactions parameters.[13] A comparison with experiment has shown that the calculated F_k values are about 20% higher while the theoretical ζ_{4f} values are about 10% overestimated. The schemes thus obtained have included energy levels up to 42,500 cm^{-1} and pertain to free-ion states.

The nonlinear hexagonal crystal NdSc$_3$(BO$_3$)$_4$ has been studied by fluorescence, absorption and excitation spectroscopy at low temperatures and RT.[14] Branching ratios, lifetimes, and experimental CF splitting of Nd^{3+} levels have been determined along with CF parameters calculated by summation of electrostatic interactions in the lattice. The spectroscopic properties of NdSc$_3$(BO$_3$)$_4$ has been compared with those of another nonlinear crystal Nd^{3+}: YAl$_3$(BO$_3$)$_4$.

Another tetraborate nonlinear crystal, Y$_x$La$_y$Sc$_z$(BO$_3$)$_4$ has been examined relative to the phases with composition $x + y + z$ = 4.[15] Large single crystals have been obtained from high-temperature solutions. The nonlinear optical coefficient d_{11} as well as the Sellmeier equations with the dispersion of the refractive indices have been determined.

The 4f electronic energy levels of Ho^{3+}: Y$_3$ (Sc, Ga)$_5$O$_{12}$ have been analyzed by means of a model Hamiltonian resulting in a RMS deviation <9 cm^{-1} between calculated and observed energy levels of Ho^{3+} ions.[16] The findings for the same crystal has been compared with those for Ho^{3+}: Y$_3$ Al$_5$O$_{12}$ and for Er^{3+}: Y$_3$Al$_5$O$_{12}$ and Er^{3+}: Y$_3$(Sc, Ga)$_5$O$_{12}$.

Energy levels of Tb^{3+} in TbAlO$_3$ have been derived from absorption transitions in the range 2940–480 nm at T = 8–300 K, as well as from high-resolution fluorescence transitions ^{5}D$_4 \rightarrow$ ^{7}F$_{6,5,4}$ at T = 85 K.[17] The experimental 58 Stark levels have been fitted with RMS deviation of 4.5 cm^{-1}. The calculated inverse magnetic susceptibility $\chi^{-1} = f(T)$ has been compared with earlier obtained experimental dependence. Also, wave functions have been obtained in the $\left| J, M_j \right>$ basis of the ground multiplet sublevels of Tb^{3+} in TbAlO$_3$ in the irreps of the C_S group. The same lanthanide complex oxide, TbAlO$_3$, has been subsequently analyzed from experimental and theoretical viewpoints.[18] By using structural data, PC and superposition models, it has been found that the low symmetry (C_s) sites of Tb^{3+} ions require a standardization procedure for the derivation of the CF parameters by means of the proposed "geometrical" rotational invariants.

In the development of detectors of ionizing radiation, a number of inorganic crystals have been studied with special emphasis on the $4f^{N-1}5d \rightarrow 4f^N$ luminescence of Ln^{3+} ions. Low-lying $4f^{N-1}5d$ excited states of Ce^{3+}, Pr^{3+}, and Tb^{3+} doped in $YAlO_3$, $Y_3Al_5O_{12}$, YPO_4, $LuPO_4$, $LaBO_3$, $YAl_3(BO_3)_4$, $Li_6Gd(BO_3)_3$ have been listed in order of increasing wavelengths.[19] An advantage of the rare earth and lanthanide complex oxides is in their very short decay times (tens of ns) at wavelength of the emission in the near UV compared to other materials. For this purpose, positions of 5d level have been compiled and analyzed for more than 300 compound containing lanthanide activators in different sites exceeding in number that figure.[20] The systematic variation of the CT energy with the number of electrons N in the $4f^{N-1}5d$ configuration has been studied in wide band gap ionic crystals including several lanthanide complex oxides.[21]

Excitation and emission spectra of YPO_4: Ln^{3+} where Ln = Ce to Yb, except Gd and Pm, have been analyzed with inclusion of the $4f^N \leftrightarrow 4f^{N-1}5d$ transitions in the regions (100–250) nm and (100–275) nm for the light and for the heavy Ln, respectively.[22,23] Thus, CF splittings arising from 5d electron have been clearly observed and a good agreement has been achieved in the comparison between experimental and simulated spectra.

In an alternative treatment of Ln ions in a free state and with $4f^{N-1}5d$ electron configurations, their spectroscopic parameters have been derived.[24] The main result of the performed computations is based on linear expressions which have considerably facilitated the derivation by reducing the number of independent parameters in the Hamiltonian.

The positions of 16 electronic energy levels of Gd^{3+} ions have been determined from absorption spectra of $LiGdP_4O_{12}$ single crystal at $T = 293$ K in the near UV range 190–340 nm.[25] FWHM for the most intense transition $^8S_{7/2} \rightarrow {}^6I_{15/2}$ at 36,680 cm^{-1} is 576 cm^{-1}. The RMS deviation between experimental and theoretical 4f energy levels has been ±118 cm^{-1}. The theoretical values of the energy levels have been obtained by means of two spectroscopic parameters only: F_2 and ζ_{4f}.

UV excitations with LEDs at six wavelengths have been applied to achieve fluorescence in flux-grown single crystals of $LiLnP_4O_{12}$, Ln = Pr, Nd, Sm, Eu, Gd, Tb, Ho, Er, Yb.[26] These excitation sources have significantly increased the number of observed 4f–4f transitions compared to the earlier studies with λ_{exc} higher than 450 nm. The assignments of the fluorescence transitions have been assisted by the energy-levels differences known from absorption spectra. The fluorescence transitions have been used to derive the 4f energy-levels diagrams of Ln^{3+} ions in these LiLnTP. Fluorescence spectrum of $LiTbP_4O_{12}$ is presented in the following figure. FWHM for the emission at 492.96 nm was obtained after excitation at 255 nm is 669 cm^{-1}.

6.4 THE BORN–HABER CYCLE REVISITED

The Born–Haber (or Born–Haber–Fajans) cycle has been applied in the evaluation for the first time of the standard changes of lattice enthalpies $\Delta_L H^\ominus$ of several Ln^{3+} complex oxides: $LnAlO_3$,[39] $Ln_3Ga_5O_{12}$,[40] $LnVO_4$,[41] $LnPO_4$,[42] $LnFeO_3$,[43] $Ln_3Fe_5O_{12}$.[44] The amount of energy per mole that binds the ions in a crystal lattice is its basic characteristics. The method is known from thermochemistry and is based on the law that the overall sum of all changes of standard enthalpies (CSEs) for a cyclic process is equal to zero. $\Delta_L H^\ominus$ can be determined following the convention[45] for signs of the quantities, symbols, including designation of phases, stoichiometry preservation, and direction of all steps involved in the cycle as physical or chemical changes. Lattice enthalpies are treated as experimental as far as the CSEs used have been experimentally obtained. Further, $\Delta_L H^\ominus$ may be compared with theoretical values derived by summation of all electrostatic interactions in a crystal, for example, those performed in[1] or with the equation of Madelung, or with any empirical equation, for example, that one of Glasser–Jenkins.[46] In most cases, the differences between the values of $\Delta_L H^\ominus$ found by the Born–Haber cycle and those from the last two equations do not exceed a few percent (Table 6.3).

TABLE 6.3 Born–Haber Cycle for Lanthanide Gallium Garnets, $Ln_3Ga_5O_{12}$.

Sr. no.	Equation of the process in each step	$\Delta H^\ominus$
1.	$Ln_3Ga_5O_{12}$ (s) → $(3/2)Ln_2O_3$(s) + $(5/2)Ga_2O_3$(s)	$-\Delta_{f, ox}H^\ominus$ $(Ln_3Ga_5O_{12})$ (s)
2.	$(3/2)Ln_2O_3$(s) + $(5/2)Ga_2O_3$(s) → 3Ln(s) + 5Ga(s) + $6O_2$(g)	$-(3/2)\,\Delta_f H^\ominus(Ln_2O_3)$(s), $-(5/2)\,\Delta_f H^\ominus(Ga_2O_3)$(s)

TABLE 6.3   *(Continued)*

3.	$3Ln(s) + 5Ga(s) + 6O_2(g) \rightarrow 3Ln(g) + 5Ga(g) +$ $6O_2(g)$	$3\Delta_s H^{\ominus}(Ln), 5\Delta_s H^{\ominus}(Ga)$
4.	$3Ln(g) + 5Ga(g) + 6O_2(g) \rightarrow 3Ln(g) + 5Ga(g) +$ $12O(g)$	$6\Delta_d H^{\ominus}(O_2)$
5.	$3Ln(g) + 5Ga(g) + 12O(g) \rightarrow$ $3Ln^{3+}(g) + 9e^- + 5Ga(g) + 12O(g)$	$3\Delta_i H^{\ominus}(Ln)$
6.	$3Ln^{3+}(g) + 9e^- + 5Ga(g) + 12O(g) \rightarrow$ $3Ln^{3+}(g) + 9e^- + 5Ga^{3+}(g) + 15e^- + 12O(g)$	$5\Delta_i H^{\ominus}(Ga)$
7.	$3Ln^{3+}(g) + 5Ga^{3+}(g) + 24e^- + 12O(g) \rightarrow 3Ln^{3+}(g) +$ $5Ga^{3+}(g) + 12O^{2-}(g)$	$12\Delta_{eg} H^{\ominus}(O)$
8.	$3Ln^{3+}(g) + 5Ga^{3+}(g) + 12O^{2-}(g) \rightarrow Ln_3Ga_5O_{12}(s)$	$-\Delta_L H^{\ominus}(Ln_3Ga_5O_{12})(s)$

The Born–Haber cycle may be presented either in graphical form (diagram) with short horizontal lines representing particular states and arrows pointing at the direction of changes, or by equations for each consecutive step. The latter is exemplified in Table 6.4.

Two effects are worth noting within the Ln series of the above complex oxides. These novelties have resulted as extensions to the lattice-energy studies and may be described in brief as follows.

(i) Plot of lattice enthalpies versus molar volumes reveals straight line with a slope $(\partial\Delta_L H^{\ominus}/\partial V_m)$ that by physical meaning, dimension

$$[\text{J mol}^{-1}]/[\text{m}^3 \text{ mol}^{-1}] = [\text{J}]/[\text{m}^3] = [\text{N m}]/[\text{m}^3] = [\text{Pa}], \qquad (6.15)$$

and values corresponds to shear modulus. All known experimental values of that modulus for various Ln series of complex oxides proved to be lower than the value of the slope $(\partial\Delta_L H^{\ominus}/\partial V_m)$. The latter may be considered as a limit of lattice energy per molar volume that, after being absorbed, would cause lattice deformation. The thermodynamic equations relating internal energy and mechanical moduli are limited to crystals of simple structure and small molar volumes.[45]

(ii) Magnetic loops within the lanthanide series of $Ln_3Fe_5O_{12}$[44] have been obtained when effective magnetic moments are expressed as a function of lattice magnetic entropies or vice versa. This effect has been observed with both experimental and theoretical values for the series of $Ln_3Fe_5O_{12}$ yielding three similar graphs.

Scheelite and zircon polymorphs of YVO_4 have been synthesized by high pressure and solid-state reaction, respectively.[47] Calculations performed by DFT have yielded structural parameters, elastic constants and bulk modulus close to the experimental values. The calculated Born effective charges for Y^{3+} in zircon YVO_4, however, are anomalously higher than the nominal charge of +3. The variations with the temperature of the heat capacity and Gibbs free energy have been also given for both polymorphs.

6.5 MICROSCOPIC SEMIEMPIRICAL MODELS

These models comprise treatments of clusters $[LnO_m]$ either in respect to spectroscopic parameters (F_k, ζ_{4f}, intensity parameters, etc.) of Ln^{3+} ions or by assessment of wave-functions overlap. Appropriate experimentally determined quantities are required in order to obtain specific quantum mechanical/chemical parameters; the needed quantities are often interionic distances, ionic radii, ε_r, F_k, ζ_{4f}, etc., as well as sets of data for comparison.

6.5.1 NEPHELAUXETIC EFFECT AND $\langle r^k \rangle_{4f}$ RRADIAL INTEGRALS OF Ln^{3+} IONS IN CRYSTALS

The dielectric screening model introduced to explain the Slater parameter shifts ΔF_k in solids[48,49] has been developed for evaluation of the same shifts by inclusion of macroscopic (ε_r) and microscopic (R_i, k, $<r^k>_{4f}$) quantities.[50] The Slater parameters F_k have been defined by eqs (6.10) and (6.11) in Section 6.3. Since,

$$\Delta F_k = F_k \text{ (free ion)} - F_k \text{ (crystal)}, \tag{6.16}$$

a precise treatment would require F_k values of free Ln^{3+} ions, that is, Ln IV. These two-particle parameters of electrostatic interactions have been known only for Pr IV,[51,52] Er IV,[53] and more recently, for Nd IV[54] and T_m IV.[55] The radial expectation values are defined as follows:

$$\left\langle r^k_{4f}(a_0)^k = \int_0^\infty r^k R_{4f^2}(r)\,dr = R_{4f}(r)|r^k|R_{4f}(r) \right\rangle \tag{6.17}$$

Thus, it has been possible to solve the reverse problem: and determine $\langle r^k \rangle_{4f}$ of Ln^{3+} ions provided that the remaining quantities of the crystals are known using the dependence

$$\left\langle r^k \right\rangle_{4f} (a_0)^k = \left\{ D_k \Delta F_k \left[\left(\frac{k}{k+1} \right) + \varepsilon_r \right] R^{2k+1} / A(\varepsilon_r - 1) \right\}^{1/2} \qquad (6.18)$$

where the coefficients D_k are those from eq (6.12), $A = (e^2/hc\text{Å}) = 1.1614 \times 10^5 \text{ cm}^{-1}$. So far, the following ions in crystals have been studied in this relation: Er^{3+},[56] Pr^{3+},[57] Nd^{3+},[58] and Tm^{3+}.[59] The problems of implementation of eq (6.13) are in finding appropriate values of the dielectric cavity in which Ln^{3+} ions are imbedded and in ΔF_k values derived from different number of parameters, that is, within different approximations or number of energy levels included in the fit. After the proposed procedure, $\langle r^k \rangle_{4f}$ of Ln^{3+} ions can be obtained using experimental data for the particular crystal. The radial expectation values thus obtained for the above Ln^{3+} ions in YAG, YPO_4, $LaAlO_3$, $M^{II}LnAlO_4$, $M^{II} = Ca$, Sr, $SrLaAlO_4$, and other oxide materials are compatible with those calculated by means of Ln^{3+} wave functions.

The related nephelauxetic ratios β_k are defined as follows:

$$\beta_k = F_k \text{ (crystal)}/F_k \text{ (free ion).} \qquad (6.19)$$

Mean nephelauxetic parameters $\bar{\beta}_k$ have been further used to determine the covalency parameter $\delta\%$ and the bonding parameter b of mixing the wave functions for the two-center $Ln^{3+}-$ oxygen bond:

$$\delta\% = [(1/\bar{\beta}_k) - 1]100, \; b^{1/2} = [(1 - \bar{\beta}_k)/2]^{1/2}. \qquad (6.20)$$

A single parameter has been proposed instead the Slater parameters F_k of Pr^{3+} in different solids.[60] The decrease of F_k (free ion) to F_k (crystal) has been assigned, as expected, to the relative dielectric permittivity of the solid and the polarizability of the ligands F^-, Cl^-, O^{2-}, and Br^-. The assumed mechanism implies indirect electrostatic interaction between the two 4f electrons with inclusion of the polarizability of the ligand and subsequently induced dominating electric dipole moment.

6.5.2 WAVE-FUNCTIONS OVERLAP IN LASING LANTHANIDE COMPLEX OXIDES

A number of TCOIs Ln^{3+}–Ln^{3+} and Ln^{3+}–O^{2-} with different axial symmetries have been computed for 10 lanthanide complex oxides.[61] The Ln^{3+} ions in these stoichiometric compounds are isolated by constituent groups AlO_6, BO_3, PO_4, WO_4, and so on, which preserve to some extent the population inversion compared to that one in doped materials. Certain crystal structures are displayed in Section 6.2. Nd^{3+} ion is the most favorable among the Ln series for lasing at RT due to its appropriate energy-level structure that allows emissions at 1.06 μm ($^4F_{3/2} \rightarrow {}^4I_{11/2}$) and at 1.32 μm ($^4F_{3/2} \rightarrow {}^4I_{13/2}$). Ground-state analytical four-term and six-term SCF wave functions of Pr^{3+} and Nd^{3+} have been compared by quantitative estimates of 4f–4f and 2v–4f (v = 2s, 2pσ, 2pπ) ion-pair overlap. In the same study, measured lifetimes of certain laser transitions of Pr^{3+}, Nd^{3+}, Ho^{3+}, and Er^{3+} in complex lanthanide oxides have been related with the TCOIs. The results of the comparative analysis emphasize the importance of this microscopic treatment for a better understanding of the physical mechanisms involved in lasing of crystals of stoichiometric compounds.

In the calculations of the TCOIs by the classical method developed by Lofthus,[62] it has been pointed out for the first time that the polynomials of incomplete gamma functions A_k and B_k are in fact coefficients reducing the power dependence of a TCOI on the distance R between the two atoms.[63] This effect has been exemplified with analytical STOs for various two centers from H_2 to UCl_3 including some title compounds.

6.6 APPLICATIONS

The use of Ln complex oxides depends on their composition, structure, particle size, and properties. The composition of the materials may vary from stoichiometric through intermediate (with partial substitution—of order of 10 mol% or higher—of Ln elements) to doped or activated (about 1 mol%). The Ln purity should be higher than 4N, especially for certain optical applications, like lasing, and the presence of other members of the Ln series should be excluded. Bulk materials, for example, single crystals, ceramics, obviously differ in applications from individual nanoparticles implemented as probes in medicine for optical or magnetic resonance imaging. Matching the crystal structure of the active layer with that of the substrate (or vice versa) has critical importance, in particular when a Ln/R complex oxide serves as layer or thin film. Properties may comprise the whole spectrum of properties: chemical, magnetic, mechanical, nuclear, optical, thermal, etc; any combination of them is directly subordinated to the requirements of the application. Areas of applications are summarized in Table 6.4.

TABLE 6.4   Certain Applications of Lanthanide/Rare-Earth Complex Oxides.

Ln/R complex oxide	Structuretype	Property	Application	References
$NdSc_3(BO_3)_4$	Hexagonal, huntite	Nonlinearoptical	Photonics	[14]
$LnAlO_3$, Ln = Sm–Yb	Perovskite, *Pbnm*	Dielectric quality factor	Dielectric resonators	[2, 64]
$LaAlO_3$	Amorphous	Dielectric, ε_r = 11.0–11.5	Thin-film transistors	[36]
$Ln_3Ga_5O_{12}$, Ln = Nd–Lu	Cubic, O_h^{10}–Ia $\bar{3}$ d	Optical, lasing	Light-emitting, stoichiometric/ doped in lasers	[65, 66]
$LnPO_4$ doped nanocrystals	Monoclinic monazite, $P2_1/n$; tetragonalxenotime, $I4_1/amd$	Multicolored upconversion	Yb^{3+}, Tm^{3+}: $LuPO_4$ (blue),Yb^{3+}, Er^{3+}:$YbPO_4$ (green)	[67]
RPO_4LnPO_4 doped nanorods/ nanowires	as above	Conductivity; thermal conductivity; adsorption; host crystal; optical, nuclear, magnetic	Protonic conductors; coatings for thermal protection; ceramic media for disposal of nuclear wastes; lasing; fluorescence and magnetic resonance imaging	[68–74]

TABLE 6.4 *(Continued)*

Ln/R complex oxide	Structuretype	Property	Application	References
LnVO$_4$ nanosheets	Zircon (ZrSiO$_4$)$I4_1/amd$	Photo sensitivity	Solar cells	[74]
YVO$_4$	Tetragonal bipyramidal D$_{4h}$	Transparency, polarizing; host crystalhost crystal	Polarizing prisms; Nd^{3+}: YVO$_4$, lasing; Eu^{3+}: YVO$_4$, red phosphor in CRTs	[75–77] 76 77
LnEO$_3$, E=Cr, Mn, Fe, Co, Ni), nanocrystalline	Orthorhombic *Pbnm*	Magnetic	Magnetic resonance imaging	[78]
Ln$_3$Fe$_5$O$_{12}$, nanocrystalline	Cubic,O$_h^{10}$–Ia$\bar3$d	Magnetic	Magnetic resonance imaging	[78]

ACKNOWLEDGMENTS

The author would like to thank Prof. B. M. Angelov for helpful discussions.

KEYWORDS

- **lanthanidecomplex oxides**
- **bonding effects**
- **lattice enthalpies**
- **magnetic**
- **dielectric properties**
- **optical spectra**

REFERENCES

1. Moretti, R.; Ottonello, G. An Appraisal of End Member Energy and Mixing Properties of Rare Earth Garnets. *Geochim. Cosmochim. Acta.* **1998**, *62* (7), 1147–1173.
2. Vasylechko, L.; Senyshin, A.; Bismayer, U. Rare Earth Aluminates and Gallates. In *Handbook on the Physics and Chemistry of Rare Earths*; Gschneidner, Jr. K. A., Bünzli, J.-C. G., Pecharski, V. K. Eds.; Elsevier: Amsterdam, 2009; pp 113–295.
3. Hong, H.Y.-P. Crystal Structure of $NdLiP_4O_{12}$. *Mat. Res. Bull.* **1975**, *10*, 635–640.
4. Hong, H.Y.-P.; Chinn, S. R. Influence of Local-Site Symmetry on Fluorescence Lifetime in High-Nd-Concentration Laser Materials. *Mat. Res. Bull.* **1976**, *11*, 461–468.
5. Lide, D. Ed. *CRC Handbook of Chemistry and Physics*, 85th ed.; CRC Press: Boca Raton, 2004; pp 4–24.
6. Tanner, P.A. Some Misconceptions Concerning the Electronic Spectra of Tri-Positive Europium and Cerium. *Chem. Soc Rev.* **2013**, *42*, 5090–5101.
7. Morrison, C. A.; Leavitt, R. P. Spectroscopic Properties of Triply Ionized Lanthanides in Transparent Host Crystals (Chapter 46). In *Handbook on the Physics and Chemistry of Rare Earths*; Gschneidner, Jr., K. A., Eyring, L. Eds., Vol. 5, North-Holland: Amsterdam, 1982; pp 461–492.
8. Condon, E. U.; Shortley, G. H. *The Theory of Atomic Spectra*; Cambridge University Press: London, 1977; pp 117–180.
9. Reisfeld, R. Optical Properties of Lanthanides in Condensed Phase, Theory and Applications. *AIMS Mater. Sci.* **2015**, *2* (2), 37–60.
10. Dieke, G. H. *Spectra and Energy Levels of Rare Earth Ions in Crystals*; John Wiley & Sons: New York, 1968; pp 63–78.
11. Wybourne, B. G. *Spectroscopic Properties of Rare Earths*; John Wiley & Sons: New York, 1965; p 15.
12. Petrov, D. N.; Angelov, B. M. Free Ions Pr IV ($4f^2$) and Tm IV ($4f^{12}$) in Intermediate versus LS Coupling Scheme. *J. Math. Chem.* **2013**, *51*, 2179–2186.
13. Ma, C.-G.; Brik, M. G.; Liu, D.-X.; Feng, B.; Tian, Ya; Suchocki, A. Energy Level Schemes for the Di-, Tri-, and Tetravalent Lanthanides and Actinides in a Free State. *J. Lumin.* **2016**, *170*, 369–374.
14. Gruber, J. B.; Reynolds, T. A.; Keszler, D. A.; Zandi, B. Spectroscopic Properties of Nonlinear $NdSc_3(BO_3)_4$. *J. Appl. Phys.* **2000**, *87* (10), 7159–7163.
15. Ye, N.; Stone-Sundberg, J. L.; Hruschka, M. A.; Aka, G.; Kong, W.; Kaszler, D. A. Nonlinear Optical Crystal $Y_xLa_ySc_z(BO_3)_4$ (x+y+z = 4). *Chem. Mater.* **2005**, *17* (10), 2687–2692.
16. Pugh, V. J.; Richardson, F. S.; Gruber, J. B.; Seltzer, M. D. Characterization and Analysis of the 4f-Electronic States of Trivalent Holmium in Yttrium Scandium Gallium Garnet. *J. Phys. Chem. Solids* **1997**, *58* (1), 85–101.
17. Gruber, J. B.; Nash, K. L.; Yow, R. M.; Sardar, D. K.; Valiev, U. V.; Uzokov, A. A.; Burdick, G. W. Spectroscopic and Magnetic Susceptibility Analyses of the ^7F_J and $^5D_4$ Energy Levels of Tb^{3+} ($4f^8$) in $TbAlO_3$. *J. Lumin.* **2008**, *128*, 1271–1284.

18. Rudowicz, C.; Gnutek, P.; Brik, V. G. Low Symmetry Aspects in Spectroscopic andMagnetic Susceptibility Studies of Tb^{3+} ($4f^8$) in $TbAlO_3$. *J. Rare Earths* **2009,** *27* (4), 627–632.

19. Dorenbos, P. The $4f^n \leftrightarrow 4f^{n-1}5d$ Transitions of the Trivalent Lanthanides in Halogenidesand Chalcogenides. *J. Lumin.* **2000,** *91,* 91–100.

20. Dorenbos, P. The 5d Level Positions of the Trivalent Lanthanides in Inorganic Compounds. *J. Lumin.* **2000,** *91,* 155–176.

21. Dorenbos, P. Systematic Behaviour in Trivalent Lanthanide Charge Transfer Energies. *J. Phys.Condens. Matter* **2003,** *15,* 8417–8434.

22. van Peterson, L.; Reid, M. F.; Wegh, R. T.; Soverna, S.; Meijerink, A. $4f^n \rightarrow 4f^{n-1}5d$ Transitions of the Light Lanthanides: Experiment and Theory. *Phys. Rev.* **2002,** *B 65,* 045113.

23. van Peterson, L.; Reid, M. F.; Burdick, G. W.; Meijerink, A. $4f^n \rightarrow 4f^{n-1}5d$ Transitions of the Heavy Lanthanides: Experiment and Theory. *Phys. Rev.* **2002,** *B 65,* 045114.

24. Ma, C.-G.; Brik, M. G.; Tian, Y.; Li, Q.-X. Systematic Analysis of Spectroscopic Characteristics of the Lanthanide and Actinide Ions with the $4f^{N-1}5d$ and $5f^{N-1}6d$ ElectronicConfigurations in a Free State. *J. Alloys Compds.* **2014,** *603,* 255–267.

25. Petrov, D.; Angelov, B. Electronic Energy Levels of Lithium Gadolinium Tetraphosphate.*Optik* **2013,** *124,* 6338–6340.

26. Petrov, D.; Eftimov, T. Fluorescence of UV-Excited $LiLnP_4O_{12}$ Single Crystals. 2016 (Unpublished Results).

27. Hatscher, S.; Schilder, H.; Lueken, H.; Urland, W. Practical Guide to Measurement and Interpretation of Magnetic Properties. *Pure Appl. Chem.* **2005,** *77* (2), 497–511.

28. Petrov, D.; Angelov, B. Preparation and Characterisation of $NdAlO_3$ Nanocrystals by Modified Sol–Gel Method. *J. Sol-gel Sci. Tech.* **2010,** *53* (2), 227–231.

29. Petrov, D.; Angelov, B.; Lovchinov, V. Sol–Gel Synthesis, Surface and Magnetic Properties of Nanocrystalline $SmAlO_3$. *J. Rare Earths* **2010,** *28* (4), 602–605.

30. Petrov, D.; Angelov, B.; Lovchinov, V. Magnetic Susceptibility and Surface Properties of $EuAlO_3$ Nanocrystals. *J. Alloys Compounds* **2011,** *509,* 5038–5041.

31. Petrov, D.; Angelov, B.; Lovchinov, V. Metamagnetic $DyAlO_3$ Nanoparticles with Very Low Magnetic Moment. *J. Sol Gel Sci. Tech.* **2011,** *58* (3), 636–641.

32. Petrov, D. Nanocrystalline $GdAlO_3$: XPS, EPR, and Magnetic Susceptibility Studies. *Appl. Phys.* A. **2011,** *104* (4), 1237–1242.

33. Anderson, P. W. Theory of Magnetic Exchange Interactions: Exchange in Insulators and Semiconductors. *Solid State Phys.* **1963,** *14,* 99–214.

34. Petrov, D.; Angelov, B. Indirect Exchange Interactions in Orthorhombic Lanthanide Aluminates. *Acta Phys. Polon.* **2012,** *4,* 737–740.

35. Bodurov, I.; Petrov, D. Dielectric Spectroscopy of Gadolinium Monoaluminate Nanoparticles. *Mater. Discov.* **2016.** http://dx.doi.org/10.1016/j.md.2015.12.002.

36. Remya, G. R.; Solomon, S.; Thomas, J. K.; John, A. Optical and Dielectric Properties of Nano $GdAlO_3$. *Mater. Today Proc.* **2015,** *2,* 1012–1016.

37. Plassmeyer, P. N.; Archila, K.; Wager, J. F.; Page, C. J.; Lanthanum Aluminum Oxide Thin-Film Dielectrics from Aqueous Solution. *ACS Appl. Mater. Interfaces* **2015,** *7* (3), 1678–1684.

38. Arun, B.; Varghese, J.; Surendran, K. P.; Sebastian, M. T. Microwave Dielectric andThermal Properties of Mixed Rare Earth Ortho Phosphate [RE$_{mix}$PO$_4$]. *Ceramics Int.* **2014**, *40*, 13075–13081.

39. Petrov, D.; Angelov, B. Lattice Energies and Polarizability Volumes of Lanthanide Monoaluminates. *Phys. B Condensed Matt.* **2012**, *407* (17), 3394–3397.

40. Petrov, D. Lattice Energies and Polarizabilities of Lanthanide Gallium Garnets (Ln$_3$Ga$_5$O$_{12}$). *Thermochim. Acta* **2013**, *557*, 20–23.

41. Petrov, D. Lattice Enthalpies of Lanthanide Orthovanadates LnVO$_4$. *Croat. Chem. Acta* **2013**, *87* (1), 85–89.

42. Petrov, D. Lattice Enthalpies, Polarizabilities and Shear Moduli of Lanthanide Orthophosphates LnPO$_4$. *Acta Chim. Slov.* **2014**, *61*, 34–38.

43. Petrov, D. Lattice Enthalpies of Lanthanide Orthoferrites LnFeO$_3$. *Acta Chim. Slov.* **2015**, *62*, 716–720.

44. Petrov, D.; Lattice Enthalpies and Magnetic Loops in Lanthanide Iron Garnets, Ln$_3$Fe$_5$O$_{12}$. *J. Chem.Thermodyn.***2015**, *87*, 136–140.

45. Atkins, P.; De Paula, J. *Physical Chemistry*, 8th ed.; Oxford University Press: Oxford, 2006; pp 50, 719, 722, 1010.

46. Glasser, L.; Jenkins, D. B. Lattice Energies and Unit Cell Volumes of Complex Ionic Solids. *J. Am. Chem. Soc.* **2000**, *122*, 632–638.

47. Huang, Z.; Zhang, L.; Pan, W.; Synthesis, Structure, Elastic Properties, Lattice Dynamics and Thermodynamics of YVO$_4$ Polymorphs from Experiments and Density Functional Theory Calculation. *J. Alloys Compounds.* **2013**, *580*, 544–549.

48. Morrison, C.; Mason, D. R.; Kikuchi, C. Modified Slater Integrals for an Ion in a Solid. *Phys. Lett.* **1967**, *A24* (11), 607–608.

49. Morrison, C. Crystal Field Parameters and Shifts of the Slater Integrals for Rare-Earth Ionsin Metals. *Phys. Lett.* **1975**, *A51* (1), 49–50.

50. Newman, D. J. Slater Parameter Shift in Substituted Lanthanide Ions. *J. Phys. Chem. Solids* **1973**, *34* (3), 541–545.

51. Sugar, J. Energy Levels of Pr^{3+} in the Vapor State. *Phys. Rev. Lett.* **1965**, *14* (18), 731–732.

52. Crosswhite, H. M.; Dieke, G. H.; Carter, W. J.; Free-Ion and Crystalline Spectra of Pr^{3+}(PrIV). *J. Chem. Phys.* **1965**, *43* (6), 2047–2054.

53. Carter, W.J.; PhD Thesis,1966, The Johns Hopkins University: Baltimore, MD; p 65.

54. Wyart, J.-F.; Meftah, A.; Tchang-Brillet, W.-Ü. L.; Champion, N.; Lamrous, O.; Spector, N.; Sugar, J. Analysis of the Free ion Nd^{3+}Spectrum (Nd IV). *J. Phys. B At. Mol.Opt. Phys.* **2007**, *40*, 3957–3972.

55. Meftah, A.; Wyart, J.-F.; Champion, N.; Tchang-Brillet, W.-Ü. L. Observation and Interpretation of the Tm^{3+}Free Ion Spectrum. *Europhys. J.* **2007**, *44*, 35–45.

56. Angelov, B. Nephelauxetic Effect and $<r^k>_{4f}$Expectation Values of Er^{3+} Ions in Ionic Solids. *J. Phys. C Solid St. Phys.***1984**, *17* (10), 1709–1712.

57. Angelov, B. Nephelauxetic Effect and $<r^k>_{4f}$ Radial Integrals of Pr^{3+}Ions in Crystals. *J. Alloy. Comp.* **2004**, *384*, 76–79.

58. Petrov, D.; Angelov, B. Radial Integrals $<r^k>_{4f}$ and Nephelauxetic Effect of Nd^{3+} in Crystals. *Spectrochim. Acta A Mol. Biomol. Spectr.* **2014**, *118*, 199–203.

59. Petrov, D. Nephelauxetic Effect and $<r^k>_{4f}$ Radial Integrals of Tm^{3+} inCrystals. *Spectrochim. Acta AMol.Biomol. Spectr.* **2015**, *151*, 415–418.

60. Tanner, P. A.; Yeung, Y. Y. Nephelauxetic Effects in the Electronic Spectra of Pr^{3+}. *J. Phys. Chem.* A **2013**, *117* (41), 10726–10735.

61. Petrov, D. Wave-Functions Overlap in Stoichiometric Lanthanide Laser Crystals. *Phys. B Condens. Matt.* **2015**, *474*, 5–8.

62. Lofthus, A. Molecular Two-Centre Integrals. I. Overlap Integrals. *Mol. Phys.* **1962**, *5* (2), 105–114.

63. Petrov, D. Polynomials for Evaluation of Two-Center Overlap Integrals. *Phys. B Condens. Matt.* **2016**, *489*, 63–66.

64. Cho, S. Y.; Kim, I.-T.; Hong, K. S. Microwave Dielectric Properties and Applications of Rare-Earth Aluminates. *J. Mater. Res.* **1999**, *14* (1), 114–119.

65. Chinn, S. R. Stoichiometric Lasers. In *CRC Handbook of Laser Science and Technology*; 1982; Vol. 1, pp 147–169.

66. Kaminskii, A. A.; Antipenko, B. M. *Multilevel Operating Schemes of Crystalline Lasers*; Nauka: Moscow,1989; p 269 (in Russian).

67. Wang, F.; Liu, X. Recent Advances in the Chemistry of Lanthanide-Doped Upconversion Nanocrystals. *Chem. Soc. Rev.* **2009**, *38*, 976–989.

68. Norby, T.; Christiansen, N. Proton Conduction in Ca and Sr-Substituted $LaPO_4$.*Solid State Ion* **1995**, *77*, 240–243.

69. Davis, J. B.; Marshall, D. B.; Oka, K. S.; Housley, P. E.; Morgan, D. Ceramic Composites for Thermal Protection Systems. *Compos.* A **1999**, *30*, 483–488.

70. Terra, O.; Dacheux, N.; Podar, R.; Clavier, N. Preparation and Characterization of Lanthanum-Gadolinium Monazites as Ceramics for Radioactive Waste Storage. *New J.Chem.* **2003**, *27*, 957–967.

71. Boatner, L. A.; Abraham, M. M.; Sales, B. C. Lanthanides Orthophosphate Ceramics for the Disposal of Actinide-Contaminated Nuclear Waste. *Inorg. Chim. Acta* **1984**, *94*, 146–148.

72. Guy, C. Audubert, F.; Lartigue, J. E.; Latrille, C.; Advocat, T.; Fillet, C. NewConditionings for Separated Long-Lived Radio Nuclides. *C. R. Phys.* **2002**, *3*, 827–837.

73. Mehta, V.; Aka, J.; Dawarb, A. L.; Mansingh, A. Optical Properties and SpectroscopicParameters of Nd^{3+}Doped Phosphate and Borate Glasses. *Opt. Mater.* **1999**, *12*, 53–63.

74. Wang, G.; Peng, Q.; Li, Y. Lanthanide-Doped Nanocrystals: Synthesis, Optical-Magnetic Properties, and Applications. *Acc. Chem. Res.* **2011**, *44* (5), 322–332.

75. Deshazer, L. G. Yttrium Orthovanadate Optical Polarizer. US Patent 3914018, 1975, 10–21. Union Carbide Corp.

76. Koechner, W. Nd^{3+}: YVO_4. *Solid-State Laser Engineering*, Springer Science+Business Media, Inc. 2006; p 69, e-ISBN : 0-387-29338-8.

77. Caro, P. Rare Earths in Luminescence. *Rare Earths* **1998**, 323–325.

78. Patil, K.C.; Hedge, M.S.; Rattan, T.; Aruna, S.T. *Chemistry of Nanocrystalline Oxide Materials*; World Scientific Publ. Co.: Singapore, 2008; pp 170, 171, 276.

CHAPTER 7

WORLD OF BIOLOGICAL ACTIVITIES AND SAFETY OF *Citrus* spp. ESSENTIAL OILS

FRANCISCO TORRENS[1,*] and GLORIA CASTELLANO[2]

[1]*Institut Universitari de Ciència Molecular, Universitat de València, Edifici d'Instituts de Paterna, POB 22085, E-46071 València, Spain*

[2]*Departamento de Ciencias Experimentales y Matemáticas, Facultad de Veterinaria y Ciencias Experimentales, Universidad Catolica de Valencia San Vicente Mártir, Guillem de Castro-94, E-46001 València, Spain*

Corresponding author. E-mail: torrens@uv.es

ABSTRACT

Citrus essential oils (EOs) are well known for flavor and fragrance properties, and numerous aromatherapeutic and medicinal applications. With the exception of phototoxicity of expressed EOs they are generally safe to use with negligible toxicity to humans. The readily available EOs will undoubtedly continue to play roles in the food and beverage industries, and for medicinal, cosmetic, and green pest-control uses. Up to 80% of the population in developing countries use traditional herbal medicine for primary healthcare. Herbal medicines play important roles in the primary healthcare of individuals and communities. However, a potential risk of unexpected drug–drug interactions exists when patients administrate herbs and are prescribed modern medication simultaneously without informing their physician of their herb use. The classical herb implicated with drug–drug interactions is St. John's wort. *Fructus aurantii* is a potential slight inducer of cytochrome P450 1A2 and 3A4, and is unlikely to impact 2E1. Caution should be paid to reduce adverse drug–drug interaction when

Fructus aurantii is administrated combined with cytochrome P450 1A2 or 3A4 substrates.

7.1 INTRODUCTION

Essential oils (EOs) and their constituents were reviewed, targeting GABAergic system and Na^+ channels as treatment of neurological diseases.[1] It was informed antifungal activity and mode of action of *Thymus vulgaris*, *Citrus limonum* (Rutaceae), *Pelargonium graveolens, Cinnamomum cassia, Ocimum basilicum,* and *Eugenia caryophyllus* EOs.[2] Composition and antioxidant potential of EO from *Citrus reticulata* fruit peels were evaluated.[3] Pomelo peel modified with citrate was reported as a sustainable adsorbent for removal of methylene blue (MB) from aqueous solution.[4] In silico docking and in vitro approaches were published toward BACE1 and cholinesterases inhibitory effect of *Citrus* flavanones.[5]

The cytochrome P450 (CYP450) enzymes, monooxygenases metabolizing xenobiotics, and endogenous substrates are involved in 80% of oxidative drug metabolism and account for 50% of the overall elimination of commonly used drugs. Among CYP450s, families 1–3 constitute 50% of total CYPs in mammals, and CYP1A2, CYP2E1, and CYP3A4 were mainly expressed in the liver, which is the major site of CYP450-mediated metabolism. The CYP450 was reported to mediate drug–drug interaction. The well-known example is that grapefruit juice increases the bioavailability of a number of drugs because of the inhibition of CYP3A4 metabolism.[6] Potential metabolic drug–drug interaction of *C. aurantium* (*Fructus aurantii, zhiqiao*) was evaluated by its effect on three CYP450s.[7]

In earlier publications in Nereis, etc., it was reported the molecular classification of yams,[8] soya bean, Spanish legumes, commercial soya bean,[9,10] fruits proximate and mineral content,[11] and food spices proximate content[12] by PCA, cluster (CA) and meta-analyses. Polyphenols,[13] flavonoids,[14] stilbenoids,[15] triterpenoids, steroids,[16] isoflavonoids,[17] natural sesquiterpene lactones,[18,19] artemisinin, and their derivatives[20–23] were analyzed. It was informed the classification of *Citrus*, principal components, cluster and meta-analyses,[24] and the extraction of natural products found in vegetal species clove and citrus.[25] The main aim of the present report is to review the biological activities and safety of *Citrus* spp. EOs.

7.2 BIOLOGICAL ACTIVITIES AND SAFETY OF *Citrus* spp. EOS

It was reviewed bioactivities and safety of *Citrus* spp. EOs: sweet orange (*Citrus sinensis*), bitter orange (*C. aurantium*), neroli (*C. aurantium*), orange petitgrain (*C. aurantium*), mandarin (*C. reticulata*), lemon (*C. limon*), lime (*C. aurantifolia*), grapefruit (*Citrus × paradisi*), bergamot (*C. bergamia*), Yuzu (*C. junos*), and kumquat (*C. japonica*).[26] Figure 7.1 shows the key volatile components.

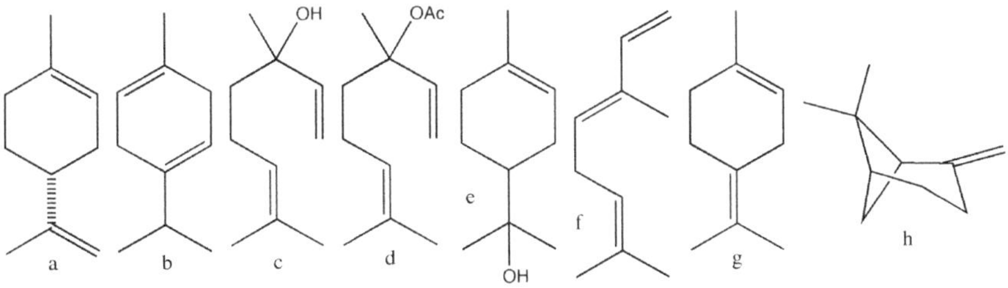

FIGURE 7.1 (a) *d*-Limonene; (b) γ-terpinene; (c) linalool; (d) linalyl acetate; (e) α-terpineol; (f) (*E*)-β-ocimene; (g) terpinolene; (h) β-pinene.

Figure 7.2 shows the key nonvolatile components.

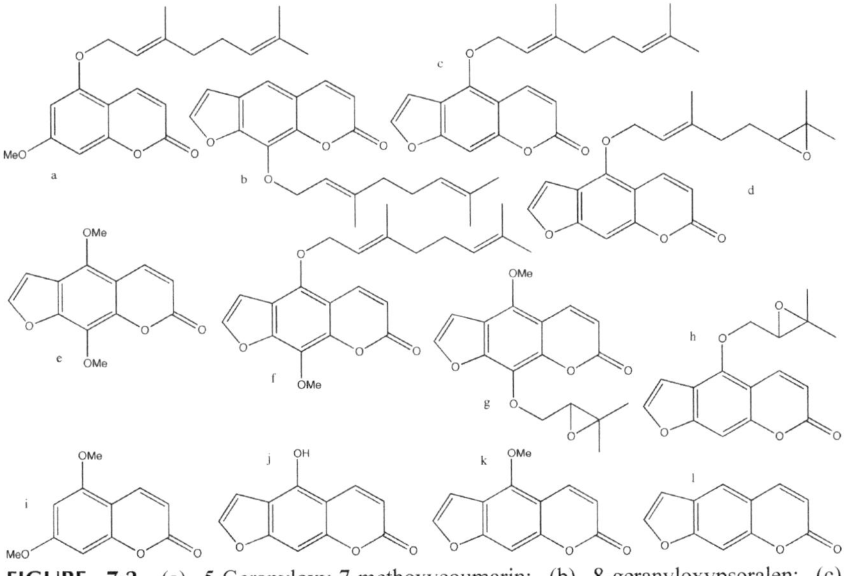

FIGURE 7.2 (a) 5-Geranyloxy-7-methoxycoumarin; (b) 8-geranyloxypsoralen; (c) bergamottin; (d) epoxybergamottin; (e) isopimpinellin; (f) 5-geranyloxy-8-methoxypsoralen; (g) byakangelicol; (h) oxypeucedanin; (i) citropten; (j) bergaptol; (k) bergapten; (l) psoralen.

7.3 DISCUSSION

Citrus EOs are well known for flavor and fragrance properties, and numerous aromatherapeutic and medicinal applications. With the exception of phototoxicity of expressed EOs, they are generally safe to use with negligible toxicity to humans. The readily available EOs will undoubtedly continue to play roles in the food and beverage industries, and for medicinal, cosmetic, and green pest-control uses.

Up to 80% of the population in developing countries use traditional herbal medicine for primary healthcare. Herbal medicines play important roles in the primary healthcare of individuals and communities. However, a potential risk of unexpected drug–drug interactions exists when patients administrate herbs and are prescribed modern medication simultaneously without informing their physician of their herb use. The classical herb implicated with drug–drug interactions is St. John's wort.[27,28]

7.4 FINAL REMARK

From the present results and discussion the following final remark can be drawn.

(1) *Fructus aurantii* is a potential slight inducer of cytochrome P450 1A2 and 3A4, and is unlikely to impact 2E1. Caution should be paid to reduce adverse drug–drug interaction when *Fructus aurantii* is administrated combined with cytochrome P450 1A2 or 3A4 substrates.

ACKNOWLEDGMENTS

The authors thank for the support from Generalitat Valenciana (Project No. PROMETEO/2016/094) and Universidad Católica de Valencia *San Vicente Mártir* (Project Nos. UCV.PRO.17-18.AIV.03 and 2019-217-001).

KEYWORDS

- **drug–drug interaction**
- **sweet orange**
- **bitter orange**
- **neroli**
- **orange petitgrain**
- **mandarin**
- **lemon**

REFERENCES

1. Wang, Z. J.; Heinbockel, T. Essential Oils and Their Constituents Targeting the Gabaergic System and Sodium Channels as Treatment of Neurological Diseases. *Molecules* **2018,** *23*, 1–24.
2. Gucwa, K.; Milewski, S.; Dymerski, T.; Szweda, P. Investigation of the Antifungal Activity and Mode of Action of *Thymus vulgaris, Citrus limonum, Pelargonium graveolens, Cinnamomum cassia, Ocimum basilicum,* and *Eugenia caryophyllus* Essential Oils. *Molecules* **2018,** *23*, 1–18.
3. Goyal, L.; Kaushal. S. Evaluation of Chemical Composition and Antioxidant Potential of Essential Oil from *Citrus reticulata* Fruit Peels. *Adv. Res.* **2018,** *15*, 1–9.
4. Ren, Y.; Cui, C.; Wang, P. Pomelo Peel Modified with Citrate as a Sustainable Adsorbent for Removal of Methylene Blue from Aqueous Solution. *Molecules* **2018,** *23*, 1–11.
5. Lee, S.; Youn, K.; Lim, G.; Lee, J.; Jun, M. In Silico Docking and In Vitro Approaches Towards Bace1 and Cholinesterases Inhibitory Effect of Citrus Flavanones. *Molecules 2018, 23*, 1509-1-12.
6. Yoshida, N.; Takagi, A.; Kitazawa, H.; Kawakami, J.; Adachi, I. Inhibition of P-glycoprotein-mediated Transport by Extracts of and Monoterpenoids Contained in Zanthoxyli Fructus. *Toxicol. Appl. Pharmacol.* **2005,** *209*, 167–173.
7. Zhou, L.; Cui, M.; Zhao, L.; Wang, D.; Tang, T.; Wang, W.; Wang, S.; Huang, H.; Qiu, X. Potential Metabolic Drug–Drug Interaction of *Citrus aurantium* L. (*Rutaceae*) Evaluating by its Effect on 3 CYP450. *Front. Pharmacol.* **2018,** *9*, 895-1-11.
8. Torrens-Zaragozá, F. Molecular Categorization of Yams by Principal Component and Cluster Analyses. *Nereis* **2013,** *5*, 41–51.
9. Torrens, F.; Castellano, G. From Asia to Mediterranean: Soya Bean, Spanish Legumes and Commercial *Soya Bean* Principal Component, Cluster and Meta-analyses. *J. Nutr. Food Sci.* **2014,** *4* (5), 98.
10. Torrens, F.; Castellano, G. Principal Component, Cluster and Meta-Analyses of Soya Bean, Spanish Legumes and Commercial Soya Bean. In *High-performance*

Materials and Engineered Chemistry; Torrens, F., Balköse, D., Thomas, S., Eds; Apple Academic–CRC: Waretown, New Jersey, 2018, pp 267–294.

11. Torrens-Zaragozá, F. Classification of Fruits Proximate and Mineral Content: Principal Component, Cluster, Meta-analyses. *Nereis* 2015, 2015(7), 39-50.

12. Torrens-Zaragozá, F. Classification of food spices by proximate content: Principal component, cluster, meta-analyses. *Nereis* **2016,** *8,* 23–33.

13. Castellano, G.; Tena, J.; Torrens, F. Classification of Polyphenolic Compounds by Chemical Structural Indicators and Its Relation to Antioxidant Properties of *Posidonia oceanica* (L.) Delile. *MATCH Commun. Math. Comput. Chem.* **2012,** *67,* 231–250.

14. Castellano, G.; González-Santander, J. L.; Lara, A.; Torrens, F. Classification of Flavonoid Compounds by Using Entropy of Information Theory. *Phytochemistry* **2013,** *93,* 182–191.

15. Castellano, G.; Lara, A.; Torrens, F. Classification of Stilbenoid Compounds by Entropy of Artificial Intelligence. *Phytochemistry* **2014,** *97,* 62–69.

16. Castellano, G.; Torrens, F. Information Entropy-based Classification of Triterpenoids and Steroids from *Ganoderma. Phytochemistry* **2015,** *116,* 305–313.

17. Castellano, G.; Torrens, F. Quantitative Structure–antioxidant Activity Models of Isoflavonoids: A Theoretical Study. *Int. J. Mol. Sci.* **2015,** *16,* 12891–12906.

18. Castellano, G.; Redondo, L.; Torrens, F. QSAR of Natural Sesquiterpene Lactones as Inhibitors of Myb-dependent Gene Expression. *Curr. Top. Med. Chem.* **2017,** *17,* 3256–3268.

19. Torrens, F.; Redondo, L.; León, A.; Castellano, G. Structure–activity Relationships of Cytotoxic Lactones as Inhibitors and Mechanisms of Action. *Curr. Drug Discov. Technol.* Submitted for Publication.

20. Torrens, F.; Redondo, L.; Castellano, G. Artemisinin: Tentative Mechanism of Action and Resistance. *Pharmaceuticals* **2017,** *10,* 20–24.

21. Torrens, F., Redondo, L.; Castellano, G. Reflections on Artemisinin, Proposed Molecular Mechanism of Bioactivity and Resistance. In *Applied Physical Chemistry with Multidisciplinary Approaches*; Haghi, A. K., Balköse, D., Thomas, S., Eds; Apple Academic–CRC: Waretown, New Jersey, 2018; pp 189–215.

22. Torrens, F.; Castellano, G. Chemical/Biological Screening Approaches to Phytopharmaceuticals. In *Research Methods and Applications in Chemical and Biological Engineering*; Pourhashemi, A., Deka, S.C., Haghi, A.K., Eds; Apple Academic–CRC: Waretown, New Jersey, In Press.

23. Torrens, F.; Castellano, F. Chemical Components from Artemisia austro-yunnanensis, Anti-inflammatory Effects and Lactones. In *Innovations in Physical Chemistry*; Haghi, A. K., Ed.; Apple Academic–CRC: Waretown, New Jersey, In press.

24. Torrens, F.; Castellano, G. Classification of Citrus: Principal Components, Cluster, and Meta-analyses. In *Applied Physical Chemistry with Multidisciplinary Approaches*; Haghi, A. K., Balköse, D., Thomas, S., Eds; Apple Academic–CRC: Waretown, New Jersey, 2018; pp 217–234.

25. Torrens, F.; Castellano, G. Extraction of Natural Products Found in Vegetal Species: Clove/Citrus. In: *Applied Food Science and Engineering with Industrial Applications*; Aguilar, C. N., Carvajal-Millan, E., Eds; Apple Academic–CRC: Waretown, New Jersey, In Press.

26. Dosoky, N. S.;. Setzer, W. N. Biological Activities and Safety of *Citrus* spp. Essential Oils. *Int. J. Mol. Sci.* **2018,** *19,* 1966-1-25.

27. Sim, S. N.; Levine, M. A. An Evaluation of Pharmacist and Health Food Store Retailer's Knowledge Regarding Potential Drug Interactions Associated with St. John's wort. *Can. J. Clin. Pharmacol.* **2010,** *17,* e57-e63.

28. Ooi, J. P.; Kuroyanagi, M.; Sulaiman, S. F.; Muhammad, T. S.; Tan, M. L. Andrographolide and 14-deoxy-11,12-didehydroandrographolide Inhibit Cytochrome P450s in HepG2 Hepatoma Cells. *Life Sci.* **2011,** *88,* 447–454.

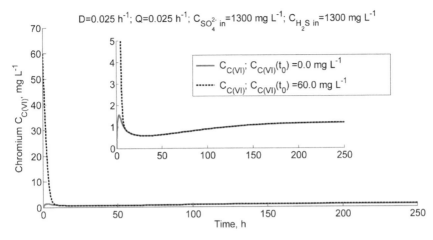

FIGURE 2.5 Dynamic behavior of the chromium variable in operating mode in the chemical reactor (CR) in continuous: open loop for two initial conditions of hexavalent chromium in CR.

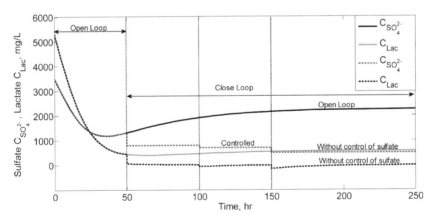

FIGURE 2.9 Dynamic behavior of the state variables in operating mode in the bioreactor (BR) in continuous: open-loop black solid line and gray solid line for sulfate and lactate, respectively, and close-loop black dashed line and red dashed line for sulfate and lactate, respectively.

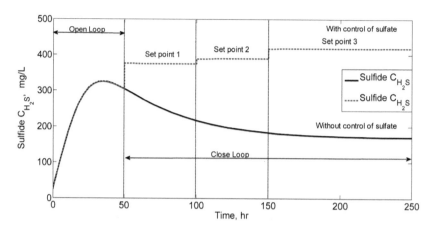

FIGURE 2.10 Dynamic behavior of the sulfide variable in operating mode in the bioreactor (BR) in continuous: open-loop black solid line and close-loop red dashed line.

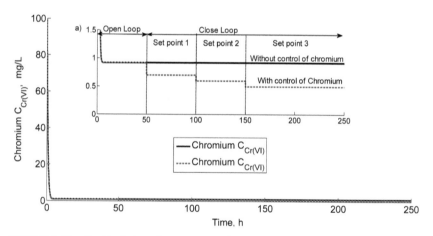

FIGURE 2.11 Residual chromium concentration.

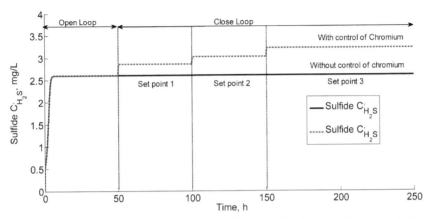

FIGURE 2.12 Dynamic behavior of the sulfide variable in operating mode in the bioreactor (CR) in continuous: open-loop black solid line and close-loop red dashed line.

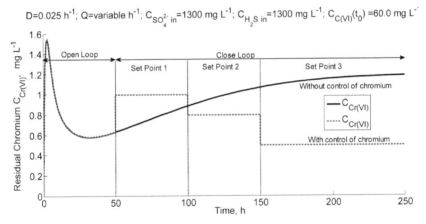

FIGURE 2.13 Dynamic behavior of the chromium variable in operating mode in the chemical reactor (CR) in continuous: open-loop black solid line and close-loop red dashed line.

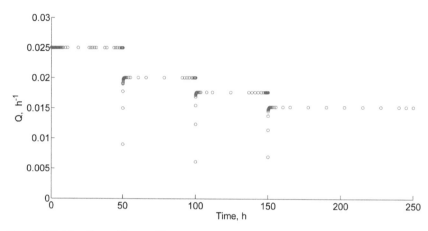

FIGURE 2.14   Control input efforts.

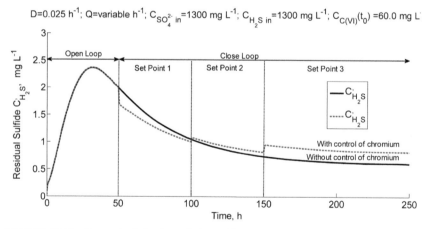

FIGURE 2.15   Dynamic behavior of the sulfide variable in operating mode in the chemical reactor (CR) in continuous: open-loop black solid line and close-loop red dashed line.

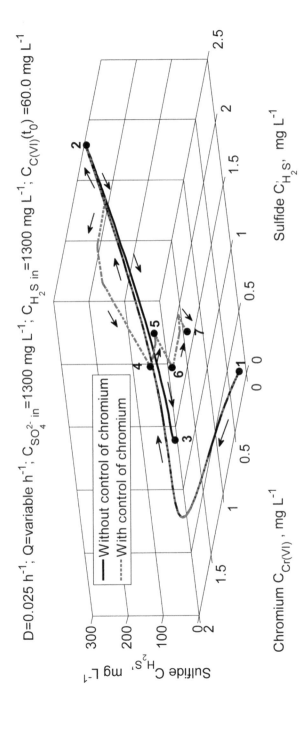

FIGURE 2.16   Open- loop and closed-loop phase portrait.

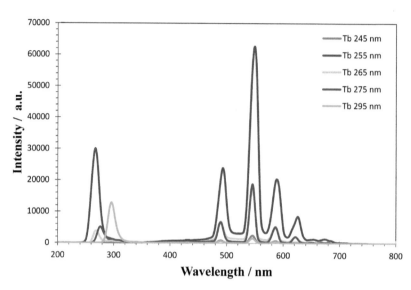

FIGURE 6.5 Fluorescence spectra of $LiTbP_4O_{12}$ single crystal at RT after excitations with LEDs between 255 and 295 nm.

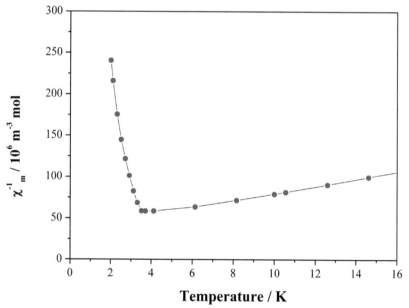

FIGURE 6.6 Variation of the inverse molar DC magnetic susceptibility $\chi_m^{-1}/10^6$ mol m^{-3} of Dy^{3+}ions in nanocrystalline DyAlO$_3$ at $T = 2$–15 K and $H = 1000$ Oe.

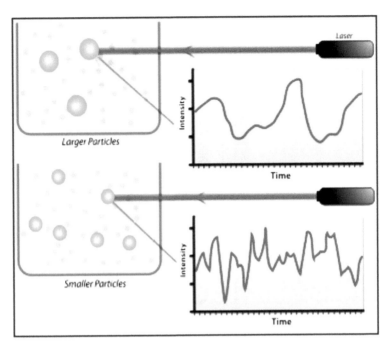

FIGURE 11.1   The principle of quasi-elastic light scattering (QLS).

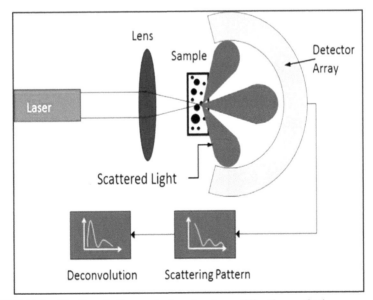

FIGURE 11.2   The Fraunhofer spectra based on laser diffraction method.

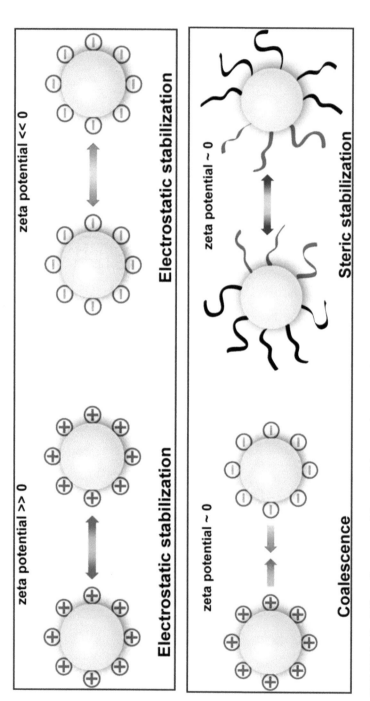

FIGURE 11.3   Effect of zeta potential on particle attraction or repulsion.

CHAPTER 8

HEALTH IN SOLITUDE AND THE THRILL OF FRAGILITY: LIVING AGREEMENT

FRANCISCO TORRENS[1,*] and GLORIA CASTELLANO[2]

[1]*Institut Universitari de Ciència Molecular, Universitat de València, Edifici d'Instituts de Paterna, POBox, E-46071 València, Spain*

[2]*Departamento de Ciencias Experimentales y Matemáticas, Facultad de Veterinaria y Ciencias Experimentales, Universidad Catolica de Valencia San Vicente Mártir, Guillem de Castro-94, E-46001 València, Spain*

Corresponding author. E-mail: torrens@uv.es

ABSTRACT

Many diseases are diseases of misery and poverty. Gálvez and coworkers informed the application of molecular topology to sustainable and environmental chemistry. Santonja organized the Patient Health Living Lab Day on Health in Solitude, thrill of fragility, and living agreement. The aim of this work is to initiate a debate by suggesting a number of questions, which can arise when addressing subjects of health in solitude, thrill of fragility, and living agreement, in different fields, and providing, when possible, answers and hypotheses. A proactive attitude is necessary: Everyone individually analyzes lifestyle and environmental impact on lifestyle/health. Health folder must include physiological/organic–emotional/cognitive–environmental–social fragilities. The subject makes clear the politics–economics–technology–science relationship. The challenge that many diseases are misery and poverty ones persists.

8.1 INTRODUCTION

Many diseases are misery and poverty ones.

Gálvez et al. reported the application of molecular topology to sustainable and environmental chemistry (Fig. 8.1).[1]

FIGURE 8.1   Molecular-topology application to sustainable and environmental chemistry. *Source:* Adapted with permission from Ateneo Mercantil de Valencia.

In earlier publications, it was informed the modeling of complex multicellular systems and tumor-immune cells competition,[2] information theoretic entropy for molecular classification of oxadiazolamines as potential therapeutic agents,[3] molecular classification of 5-amino-2-aroylquinolines and 4-aroyl-6,7,8-trimethoxyquinolines as highly potent tubulin polymerization inhibitors,[4] polyphenolic phytochemicals in cancer prevention and therapy, bioavailability versus bioefficacy,[5] molecular classification of antitubulin agents with indole ring binding at colchicine-binding site,[6] and molecular classification of 2-phenylindole-3-carbaldehydes as potential antimitotic agents in human breast cancer cells.[7] The aim of this work is to initiate a debate by suggesting a number of questions (Q), which can arise when addressing subjects of health in solitude, thrill of fragility, and living agreement, in different fields, and providing, when possible, answers (A) and hypotheses (H).

8.2 HEALTH IN SOLITUDE AND THRILL OF FRAGILITY: LIVING AGREEMENT

Santonja organized Patient Health Living Lab (PHLL) Day on Health in Solitude.[8] Todoli raised questions (Qs) on talking about treatment.[9]

Q1. Doctor, what would you do if she were your mother?

Q2. Faced with the suspicions of overdiagnosis, what questions might one ask?

Q3. Has it to do with measuring a risk factor or a symptom?

Q4. Why would they have to do for me?

Q5. Why even do they there?

Q6. When?

He raised the following additional questions.

Q7. How is my patient?

Q8. How does he live together with his disease?

Q9. How does he face his discapacities?

Q10. How to evaluate the global prognosis of the patient?

Q11. How to communicate it to the patient?

Q12. How to maintain the assistential continuity of prescription?

De la Cueva proposed hypotheses (Hs)/Qs on scientific and experiential evidences.[10]

H1. It is necessary that the patient gets involved in research.

H2. Untrue H: Science = Scientific Evidence = Experimental Evidence.

H3. For science, accord is irrelevant.

Q13. In the experiential ambit, the question is: Who has said that?

Q14. In the science ambit, the question is: How is that known?

H4. Consequences of the separation: Science loses contact with the reality, for example, Ig Nobel Prizes.

H5. (José de Letamendi). Who knows only medicine, he knows nor medicine.

Sales proposed Q/H/answers (A) on what people talk about when they talk about environmental fragility.[11]

Q15. What are people talking about when they talk about environmental fragility?

Q16. Are people exposed to the risk of failure of their fragile relationship with the environment?

H6. Lifestyle → Health.

Q17. How the state of people's environment affects them and their environmental health?

Q18. How to become aware of all that?

H7. Lifestyle → Positive and negative environmental aspects/impacts → Health.

Q19. How to determine the degree of honesty of the sources?

A19. It is difficult, but knowledge is needed to determine the degree of honesty of the sources.

H8. Air quality (AQ) passive information → Knowledge → Conducts change → AQ improvement.

Q20. What indicators can serve one to know his living agreement (LA)?

A20. Exposition times; pollution amount; how much energy/raw materials one consumes day in day out.

Q21. How much energy does one consume day in day out?

Q22. How much raw materials does one consume day in day out?

Q23. Reactive attitude?

A23. No, one should act: a proactive attitude.

She concluded for proactive attitude: Every one individually analyze lifestyle and environmental impact on lifestyle/health.

Royo proposed Q/A on what people talk about when talking about physiological/organic fragility.[12]

Q24. What are people talking about when they are talking about physiological and organic fragility?

Q25. What happens when they tell you that you suffer from diabetes?

A25. You should inject insulin into yourself.

Q26. What is it changing?

A26. Social networks.

Q27. Why are patients in social networks?

A27. The most useful themes have not to do with consultation.

Q28. What do patients need?

A28. Help, emotional support, learning, and updating.

Rocher raised questions on roadmap to health in solitude and living agreement.[13]

Q29. What can be learned from global social responsibility for health?[14]

Q30. Responsibility for health: What/concept?

Q31. How to reach the strategy?

Q32. What to do to encourage the responsibility for health?

Q33. Joint document: agreement?

Penalva proposed the following hypothesis on living agreement and health folder (HF).[15]

H9.   Triple demand: $HF \times LA = P^3$ (practicality, portability, proactivity).

She concluded HF must include physiological/organic–emotional/cognitive–environmental–social fragilities.

8.3   DISCUSSION

Many diseases are misery and poverty ones.

   A proactive attitude is necessary. Every one individually analyze lifestyle and environmental impact on lifestyle/health. The HF must include physiological/organic–emotional/cognitive–environmental–social fragilities.

   The subject makes clear the politics–economics–technology–science relationship. The challenge that many diseases are misery and poverty ones persists.

8.4   FINAL REMARKS

From the present results and discussion, the following final remarks can be drawn:

   (1) A proactive attitude is necessary. Everyone individually analyzes lifestyle and environmental impact on lifestyle/health.

(2) HF must include physiological/organic–emotional/cognitive–environmental–social fragilities.

(3) The subject makes clear the politics–economics–technology–science relationship.

(4) The challenge that many diseases are misery and poverty ones persists.

ACKNOWLEDGMENTS

The authors thank support from Generalitat Valenciana (Project No. PROMETEO/2016/094) and Universidad Católica de Valencia San Vicente Mártir (Project Nos. UCV.PRO.17-18.AIV.03 and 2019-217-001).

KEYWORDS

- molecular topology
- sustainable chemistry
- environmental chemistry
- misery disease
- poverty disease
- politics–economics–technology–science
- treatment

REFERENCES

1. Wang, J.; Land, D.; Ono, K.; Gálvez, J.; Zhao, W.; Vempati, P.; Steele, J. W.; Cheng, A.; Yamada, M.; Levine, S.; Mazzola, P.; Pasineti, G. M. Molecular Topology as Novel Strategy for Discovery of Drugs with Aβ Lowering and Anti-aggregation Dual Activities for Alzheimer's Disease. *PLoS One* **2014,** *9,* e92750–1-7.

2. Torrens, F; Castellano, G. Modelling of Complex Multicellular Systems: Tumour–Immune Cells Competition. *Chem. Central J.* **2009,** *3* (I), 75–1-1.

3. Torrens, F.; Castellano, G. Information Theoretic Entropy for Molecular Classification: Oxadiazolamines as Potential Therapeutic Agents. *Curr. Comput. Aided Drug Des.* **2013,** *9,* 241–253.

4. Torrens, F.; Castellano, G. Molecular Classification of 5-amino-2-aroylquinolines and 4-aroyl-6,7,8-trimethoxyquinolines as Highly Potent Tubulin Polymerization Inhibitors. *Int. J. Chemoinf. Chem. Eng.* **2013**, *3* (2), 1–26.

5. Estrela, J. M.; Mena, S.; Obrador, E.; Benlloch, M.; Castellano, G.; Salvador, R.; Dellinger, R. W. Polyphenolic Phytochemicals in Cancer Prevention and Therapy: Bioavailability Versus Bioefficacy. *J. Med. Chem.* **2017**, *60*, 9413–9436.

6. Torrens, F.; Castellano, G. Molecular Classification of Antitubulin Agents with Indole Ring Binding at Colchicine-Binding Site. In *Molecular Insight of Drug Design*; Parikesit, A. A., Ed.; InTechOpen: Vienna, 2018; pp 47–67.

7. Torrens, F.; Castellano, G. Molecular Classification of 2-Phenylindole-3-carbaldehydes as Potential Antimitotic Agents in Human Breast Cancer Cells. In *Theoretical Models and Experimental Approaches in Physical Chemistry: Research Methodology and Practical Methods*; Haghi, A. K.; Thomas, S.; Praveen, K. M.; Pai, A. R., Eds.; Apple Academic–CRC: Waretown, NJ, In press.

8. Santonja, F. J., Ed.; *Book of Abstracts, Health in Solitude.* Patient Health Living Lab: València, Spain, 2018.

9. Todoli, J. *Book of Abstracts, Health in Solitude.*, Patient Health Living Lab: València, Spain, 2018, 0–3.

10. De la Cueva, E. *Book of Abstracts, Health in Solitude.* Patient Health Living Lab: València, Spain, 2018; 0–4.

11. Sales, M. J. *Book of Abstracts, Health in Solitude.* Patient Health Living Lab: València, Spain, 2018, 0–5.

12. Royo, D. *Book of Abstracts, Health in Solitude.* Patient Health Living Lab: València, Spain, 2018, 0–5.

13. Rocher, E. *Book of Abstracts, Health in Solitude.* Patient Health Living Lab: València, Spain, 2018, 0–6.

14. March, J. M.; Sánchez, A., Eds. *Retos de la Transición Económica Rusa.* Universitat de València: València, Spain, 2003.

15. Penalva, C. *Book of Abstracts, Health in Solitude.* Patient Health Living Lab: València, Spain, 2018, 0–6.

CHAPTER 9

CANCER AND HYPOTHESES ON CANCER

FRANCISCO TORRENS[1,*] and GLORIA CASTELLANO[2]

[1]*Institut Universitari de Ciència Molecular, Universitat de València, Edifici d'Instituts de Paterna, POB 22085, E-46071 València, Spain*

[2]*Departamento de Ciencias Experimentales y Matemáticas, Facultad de Veterinaria y Ciencias Experimentales, Universidad Catolica de Valencia San Vicente Mártir, Guillem de Castro-94, E-46001 València, Spain*

Corresponding author. E-mail: torrens@uv.es

ABSTRACT

Mukherjee reviewed cancer. Barillot et al. revised computational systems biology of cancer. Kuang et al. introduced mathematical oncology. Complex multicellular systems and tumor-immune cells competition were modeled. A hypothesis explained how human immunodeficiency virus/acquired immunodeficiency syndrome destroy immune defences. Another hypothesis justified 2014 emergence, spread, and uncontrolled Ebola outbreak. The aim of this work is to initiate a debate by suggesting a number of questions, which can arise when addressing subjects of cancer and hypotheses on cancer, in different fields, and providing, when possible, answers and hypotheses.

9.1 INTRODUCTION

Mukherjee reviewed cancer.[1] Barillot et al. revised computational systems biology of cancer.[2] Kuang et al. introduced mathematical oncology.[3]

In earlier publications, it was informed the modeling of complex multicellular systems and tumor-immune cells competition,[4] information theoretic entropy for molecular classification of oxadiazolamines as potential therapeutic agents,[5] molecular classification of 5-amino-2-aroylquinolines and 4-aroyl-6,7,8-trimethoxyquinolines as highly potent tubulin polymerization inhibitors,[6] polyphenolic phytochemicals in cancer prevention and therapy, bioavailability versus bioefficacy,[7] molecular classification of antitubulin agents with indole ring binding at colchicine-binding site,[8] and molecular classification of 2-phenylindole-3-carbaldehydes as potential antimitotic agents in human breast cancer cells.[9] It was reported how human immunodeficiency virus/acquired immunodeficiency syndrome (HIV/AIDS) destroy immune defences, hypothesis,[10] 2014 emergence, spread, uncontrolled Ebola outbreak,[11,12] and Ebola virus disease, questions, ideas, hypotheses, and models.[13] The aim of this work is to initiate a debate by suggesting a number of questions (Q), which can arise when addressing subjects of cancer and hypotheses on cancer, in different fields, and providing, when possible, answers (A) and hypotheses (H).

9.2   HYPOTHESES ON CANCER: THE EXAMPLE OF LUNG CANCER AND SMOKING

Mukherjee reviewed and proposed the following hypotheses (Hs) on cancer. Koch (1884) stipulated that for an agent to be defined as a disease cause, it needs to fulfil three criteria:

H1.   The causal agent had to be present in diseased animals.
H2.   It had to be isolated from diseased animals.
H3.   It had to be capable of transmitting the disease when introduced into a secondary host.

By the early 1950s, cancer researchers split into three feuding camps:

H4.   Virologists (Rous): Viruses caused cancer, although no such virus was found in human studies.
H5.   Epidemiologists (Doll–Hill): Exogenous chemicals caused cancer, although no mechanistic explanation was offered.
H6.   Boveri's successors: Genes internal to the cell cause cancer.

Hill's epidemiology criteria showed nine lung cancer–smoking association additional features:[14]

H7. It was strong: the increased risk of cancer was nearly five- or 10-fold in smokers.

H8. It was consistent: Doll–Hill's and Wynder–Graham's studies, in vastly different contexts on vastly different populations, came up with the same link.

H9. It was specific: Tobacco was linked to lung cancer (site where tobacco smoke enters the body).

H10. It was temporal: Doll–Hill found that the longer one smoked, the greater the increase in risk.

H11. It possessed a biogradient: the more one smoked in quantity, the greater risk for lung cancer.

H12. It was plausible: mechanistic link between inhaled carcinogen and malignant lung change was not implausible.

H13/14. It was coherent; it was backed by experimental evidence: epidemiological and laboratory findings, for example, Graham's tar-painting experiments in mice were concordant.

H15. It behaved similarly in analogous situations: smoking was correlated with lung, and lip, throat, tongue, and oesophageal cancer.

Hanahan and Weinberg summarized the six hallmarks of cancer rules:[15,16]

H16. Self-sufficiency in growth signals: cancer cells acquire autonomous drive to proliferate (pathological mitosis) by oncogenes (*ras*, *myc*) activation.

H17. Insensitivity to growth-inhibitory (antigrowth) signals: cancer cells inactivate tumor suppressor genes (retinoblastoma, *Rb*) that normally inhibit growth.

H18. Evasion of programed cell death (apoptosis): cancer cells suppress/inactivate genes/pathways that normally enable cells to die.

H19. Limitless replicative potential: cancer cells activate specific gene pathways that render them immortal after generations of growth.

H20. Sustained angiogenesis: cancer cells acquire capacity to draw out their own supply of blood/blood vessels (tumor angiogenesis).

H21. Tissue invasion and metastasis: cancer cells acquire capacity to migrate to other organs, invade other tissues, and colonize the organs, resulting in spread throughout body.

The six-degrees-of-separation-from-cancer rule follows:

H22. One asks any biological question, no matter how seemingly distant, and will end up, in <6 genetic steps, connecting with a proto-onco-gene/tumor suppressor.

The three new directions for cancer medicine follow:

H23. Cancer therapeutics.

H24. Cancer prevention.

H25. To integrate aberrant genes/pathways understanding to explain cancer behavior as a whole.

Cancer prevention showed two controversies in epidemiology.

H26. Oestrogen/progesterone are major risk factors for incidence/fatality from oestrogen-positive breast cancer.

H27. The indiscriminate overuse of pesticides is partially responsible for rising incidence of cancer.

9.3 CANCER

Mukherjee proposed the following Qs, As, and H on cancer.

Q1.   Why did you decide to write a book about cancer?

A1.   The book is a long answer to a Q first posed to me by a patient.

Q2.   What exactly is cancer?

A2.   Cancer is not a disease but a whole family of diseases.

Q3.   Had you a particular audience in mind while writing the book?

Q4.   Was the idea to write it either for patients or for a layperson to understand?

A4.   It is written for a layperson to understand but I wanted to treat audience with utmost seriousness.

Q5.   What did made you decide to focus so much on suffering patients in the past of cancer's story?

A5.   There are other people who gave up their lives to help us understand more about this disease.

Q6.   Would you say *The Emperor of All Maladies* attests to a lack of progress in oncology?

A6.   Absolutely not, there is a clear place that lies between nihilism and overoptimism (or hype).

Q7.   Do you think one needs to change way one educates patients/public about cancer to move away from mindset that cancer is a single disease, to explain that it comprises many diseases?

A7.   Yes, I attempt to allow public to understand complexity level and appreciate ingenuity/resilience of knowledge that came about in discovery terms.

Q8.   Do you agree with Weinberg or do you feel keeping focus on genetics offers best opportunities for care?

A8.   Genetics is a vital part but only a small part of it, one piece of a much larger puzzle.

Q9.   How does every area cast illness in its own image?

Q10. What is the understanding of epigenetics in cancer?

Q11. What is the relationship between the biology of cancer and the stem cell?

Q12. How do some physicians become indifferent, not just to death but also to life?

Q13. What do you mean?

H1.   Untrue H: I do not want to become an oncologist because everyone dies.

Q14. When you must deliver bad news, how do you prepare for that?

A14. What really helps is to listen to the person you are delivering bad news to.

Q15. How may or may not they be achieved?

Q16. What is achievable?

Q17. What is not achievable?

Q18. What does it mean bad news?

Q19. How did ego/distrust between professional in different disciplines (e.g., surgery, chemotherapy) hinder progress in cancer research?

Q20. Has this got better?

A20. It got vastly better because the prior years were humbling and disciplines became less isolated.

Q21. What was happening not only medically to person but also socially, emotionally, etc., and made a valuable ally in treating a patient?

Q22. Do you think community oncologists are reluctant to adopt new discoveries in practice?

A22. No, I do not think so, I think community oncologists are really cancer-medicine frontline.

Q23. Do you think historical memory of early experiences with basically unregulated clinical trials is at all responsible for negative regard Americans today have of clinical trials and their reluctance to participate in clinical trials?

A23. A lot of reluctance exists about clinical trials because people did a bad job of educating public on what a clinical trial means, how important it is and how only way to learn with cancer is to participate.

Q24. How did people (e.g., Mary and Albert Lasker, Farber) convince public to pay attention to cancer and mobilize funding?

Q25. How do you reconcile researchers and advocates?

A25. You balance the issue via the tried-and-tested mechanisms of politics.

Q26. What is known?

Q27. Should mammography be performed between the ages of 40 and 50?

Q28. Is mammography a preventive mechanism that allows saving lives?

Q29. Does mammography save lives?

A29. Thinking deeply about what data show and making modifications in response until a compromise is reached between advocates and regulatory bodies.

Q30. How did you choose what to include in the book and what to exclude?

A30. I used a few simple criteria: a discovery in cancer biology had to transform into a medical reality.

Q31. What are areas of cancer biology where laboratory advances are becoming clinical realities?

A31. Four areas exist: immune system, cancer metabolism, gene regulation, and microenvironment.

Q32. However, what was the precise mechanism of attack?

Q33. Why were only certain cancers being attacked?

Q34. Could such an immune activation be used as a therapeutic tool?

Q35. Why do certain leukemias grow only within the bone marrow and spleen?

Q36. Why does prostate cancer metastasize to the bone?

Q37. What is link between the unique environments and tumor growth or its ability to resist drugs?

Q38. Are there specific safe harbors for certain cancer cells and might disrupting the harbors allow for therapies?

Q39. However, what is about the mounting costs of such new therapies?

Q40. As a society, can people justify and afford the escalating costs of cancer drugs?

A40. People must find a middle ground between cost and price, and they are nowhere close to it.

Q41. Effectiveness Q. Is spending $100,000 on a drug that extends lives by 8 weeks worthwhile?

Q42. Who is asking?

Q43. Quality of life Q. What is the quality of life?

Q44. Your book focuses on cancer in United States. What about cancer in the international context?

A44. Stories in the book take people to Germany, Austria, Egypt, Greece, and the United Kingdom.

Q45. What was truly distinctive about Farber?

A45. It was his role in the war on cancer.

Q46. How is cancer being tackled in other parts of the world, particularly in developing nations?

Q47. Why, aside from tobacco, asbestos, radiation, etc., do you not speak at length about other cancer prevention mechanisms?

A47. Cancer prevention is a complex issue.

Q48. How will people identify carcinogens in the future?

Q49. Can you give people more specifics about how such an accelerator or brake gene might work?

A49. List of oncogenes and tumor suppressor genes is huge (>100) and they are specific for every cancer type.

Q50. What is about the role of the mind/brain in cancer?

A50. The mind/brain connection plays an important role in one's psychic response to any illness.

Q51. Why should there be an archetypal patient?

Q52. What about alternative medicine?

A52. All medicine is alternative before it becomes mainstream.

Q53. How to use all chemicals in plants?

Q54. Is there a cancer prevention lifestyle?

Q55. As a practicing oncologist and a father, where did you find time to write a book so big/complex?

A55. What was important was to have reason to do it: trying to answer my patient's Q.

Q56. Where did mammography leave people in 1986?

9.4 WHERE NEXT-GENERATION SEQUENCING LEADS: CANCER GENETIC COMPLEXITY

Next-generation sequencing (NGS) revealed insights into genetic nature of different cancers leading to better cell culture models for disease research.[17] Because of genetic and phenotypic variability between cancers, identifying mutations that drive disease initiation and development is not easy.[18] In large-scale studies [e.g., The Cancer Genome Atlas (TCGA) Net, Catalogue of Somatic Mutations in Cancer (COSMIC) by Sanger Institute], researchers explored genetic changes in thousands of human normal/tumor paired tissue samples looking for such mutations.[19,20] Investigation revealed genetic complexity and tumor heterogeneity within a tumor from a single patient.

9.5 GENETIC ADVICE IN CANCER IN THE VALENCIA COMMUNITY (SPAIN)

Salas reported genetic advice in cancer in Valencia Community.[21] Molecular analysis showed that cancer is sporadic 70–80%, family 10–25%, and hereditary 5–10%. Hereditary cancer syndromes resulted breast/ovary (60%), hereditary nonpolyposis colorectal cancer (HNPCC, 31%), familial adenomatous polyposis (FAP), multiple endocrine neoplasia type-2 (MEN2), von Hippel–Lindau (VHL) syndrome, retinoblastoma (Rb), Peutz–Jeghers syndrome (PJS), etc.

9.6   MULTI-SCALE MODELING APPROACHES IN SYSTEMS BIOLOGY: CANCER

Genomic activity ignored complexity and dynamical aspects associated with homeostasis, central concept in physiology/medicine.[22] Emulating reductionism, major thrust was imparted to find out and rectify singular factor associated with a disease, which approach produces success toward the development of modern medicine; however, it does not fulfil expectations in investment terms, which were made to solve different complex illnesses, for example, cancer, which actually drags scientific community to adopt systems approach and biology was accepted. Mathematical-modeling objective is not to generate a large-scale computer simulation of a bioprocess, for example, metastasis, although such a model is feasible, its complexity would make it so sensitive to assumptions as to be of no practical value.[23] Most effective model presents a smaller number of variables.

9.7   MATHEMATICAL MODELING FOR CANCER DRUG DISCOVERY/DEVELOPMENT

Mathematical modeling enables in silico cancers classification, disease-outcomes prediction, therapy optimization, identification of promising drug targets, and prediction of anticancer-drugs resistance. In silico prescreened drug targets are validated by a small number of carefully selected experiments. Mathematical-modeling basics in cancer drug discovery/development was discussed. Topics included in silico discovery of novel molecular drug targets, immunotherapies optimization, personalized medicine, and guiding pre/clinical trials. Breast cancer (BC) showed mathematical-modeling applications in cancer diagnostics, identification of high-risk population, cancer screening strategies, tumor-growth prediction, and guiding cancer treatment. Mathematical models are essential components of toolkit used to fight against cancer. Combinatorial complexity of drugs discovery is enormous, making systematic medicine innovation by experimentation alone difficult if not impossible. Challenges include seamless integration of growing data, information and knowledge, making them available for analyses multiplicity. Mathematical models are essential for bringing cancer drug discovery into 'omics era, big data (BD), and personalized medicine.

9.8   USING COMPLEXITY ANALYSIS TO FIND COHERENCE IN CANCER BIG DATA

Multidimensional molecular characterization led to a tsunami of cancer data.[24] Precision medicine assumes that understanding and better interventions will flow from the BD. However, data cacophony challenged many founding cancer paradigms and, when viewed via the simplifying lenses, appeared chaotic. Cancer was observed to not only complicated but also complex. Coherence emerged from the apparent disorder when the molecular cancer BD was examined via methods of complex adaptive systems, for example, complex phenotypes (susceptibility to cancer, response to interventions) had regular, reproducible patterns when the interaction between components was taken into account via molecular nets, which examination offered a means to capture gene-centric concepts and complex, emergent behavior. In breast and liver cancers, nets provided much stronger signals of disease susceptibility than individual constitutional variants or expression of individual-genes collections. The analyses showed emergent, higher-level association of gene variation.

7.9   PERSONALIZED MEDICINE: BC TREATMENT/ APPROACHING ADVANCES

Lluch proposed Qs and As on personalized medicine and strategies in the treatment of BC.[25]

Q1.   What would have happened if Trastuzumab were classically developed?

A1.   There would not have been improvement without previous molecular diagnostic.

Q2. What is new with drug development?

Q3. What disciplines are improved in personalized medicine for cancer?

A3. All disciplines, especially bioinformation technologies.

Q4. Do you think that fighting versus cancer has to do with cancer-cell knowledge and that, when people know it, they could eliminate chemotherapy?

A4. Yes, with knowing it, interactions between these cells and pathways/ interactions between them.

She proposed the following questions and answers on advances in approaching BC:

Q5. What is a tumor?

Q6. What is a cancerous cell?

A6. Anarchic cell that multiplies without control/does not pay attention to organism calls to order.

Q7. How is a cancer born?

A7. Normal cell (mutations, oncogenes)→ Cancerous cell (second stage, 2–8 years)→ Tumor.

Q8. What proportions of BC and ovary cancer (OC) are hereditary [by BC (BRCA)1/2 genes]?

A8. An 80% BC and 40% OC are hereditary (by mutation carriers in BRCA1/2 genes).

She proposed Qs and As on new advances in approaching BC and personalized medicine:

Q9. What is a tumor?

Q10. What is a cancerous cell?

A10. An anarchic cell that multiplies without control and does not listen the organism calls to order.

Q11. Why me?

A11. People do not know it.

Q12. How does a cancer rise?

Q13. What rate of BC and ovary cancer is hereditary?

A13. The BC: 5–10%; OC: 5–10%.

Q14. Where is genomic platform taking people?

Q15. What to do in the case of intermediate risk: endocrine therapy/ chemotherapy benefit?

Q16. What disciplines are involved in personalized medicine for cancer?

9.10 WHEN RELATIVES STAY TOO LONG: IMPLICATIONS FOR CANCER MEDICINE

Maley group proposed Qs/Hs/problem (P)/A on relatives staying long and cancer implications.[26]

Q1. What is the extent of intratumor heterogeneity?

Q2. How does treatment affect clonal evolution of cancer?

H1. (Roche-Lestienne, 2003; Downing, 2008; Oliner, 2012; Stegmeier, 2015). Genetic abnormalities contributing to recurrence are selected for during treatment.

P1. Identification of trunk (clonal) mutations of a specific tumor type is a challenging problem.

H2. Relapse samples have transversions increased proportion due to DD induced by radiation therapy.

H3. Standing subclonal variation in a tumor is main factor for evolution of treatment acquired resistance.

Q3. How could single therapeutic agents be combined in order to control/ suppress clonal evolution?

H4. A resistant clone is already present at diagnosis.

Q4. How can one deal with standing variation in tumor at diagnosis and resulting evolvability?

A4. To combine drugs in hope that no cancer cell be present that be resistant to all drugs in cocktail.

Q5. (Maley, 2015). Can oncology recapitulate palaeontology?

H5. (Maley, 2015). One slows/avoids resistance evolution via cytostatic over cytotoxic drugs targeting tumor evolvability lowering mutation rate and extending generation time of cancer cells, attempting to maintain tumor control keeping sensitive cells alive (adaptive therapy) or selecting versus resistant clones.

9.11   TRANSLATING RESEARCH: NEW CHALLENGES FACED WITH PRECISION ONCOLOGY

Calabuig Fariñas proposed Qs/A on translating research and challenges in precision oncology.[27]

Q1.   How many genes are mutated in a human cancer?

A1.   For instance, 100–200 genes in lung cancer, so there is need for developing biomarkers.

Q2.   Am I in an increased risk of cancer?

Q3.   What alterations is one searching for?

Q4.   Must one search for such alterations in the beginning, mid, or end of the cascade?

Q5.   Biomarkers in cancer, where to search for?

Q6.   Analysis of biomarkers, what genes is one going to analyze?

Q7.   How many genes is one going to analyze?

Q8.   The NGS, what type of information does it generate?

Q9.   How to inform the findings associated with the search?

9.12   MATHEMATICAL ONCOLOGY: A QUANTITATIVE THEORY OF ONCOLOGY

Kuang et al. proposed Hs, Qs, answer (A) and fact (F) on a quantitative theory of oncology.

H1.   Medicine largely lacks coherent, quantitative (or better, mathematical) theories.

Q1.   How can one say such a thing after so much medical theory couched in mathematics?

A1.   Scientific medicine simply has not produced mathematical theory applicable to broad areas of medicine.

H2.   There is no sort of grand unified medical theory.

H3.   Mathematical models are applied in an ad hoc manner to biological and physical phenomena relevant to medical oncology.

H4. Deepening the connections among traditionally disjointed theoretical constructs will increase the power theory and will inform medical art.

H5. (Hanahan, Weinberg, 2000, 2011): Cancer hallmarks are phenotypes caused by evolution.

H6. Such theory is completely coherent, highly successful, becoming increasingly comprehensive, and yet devoid of mathematics.

H7. From a more general perspective, medicine looks toward genomics for its theory.

H8. To argue that genomics is not a sufficient source of coherent theory is to argue versus dominant intellectual tradition of 21st-century biology.

H9. Genomics is the opposite of theory.

H10. Theory identifies patterns within chaos.

H11. Genomics generates chaos.

H12. General genomics program, as applied to a given species, generates a consensus sequence for entire genome of that species, characterizes variation in homologous deoxyribonucleic acid (DNA) sequences among individuals, and identifies functions of functional DNA sequences.

H13. Untrue H. A gene occupies a specific locus in the genome.

H14. Genomics was successful in forcing people to reorganize their thinking, that is, build theory to explain new observations cascade.

H15. Such theory was by and large ad hoc and essentially qualitative.

F1. Detailed modeling of genomics chaos was never, in the history of science, the way forward.

Q2. How are systems to be integrated into simple patterns, common themes, understandable descriptions and explanations that fit together into a more-or-less comprehensive narrative?

H16. Genomics and systems biology programs have immense value.

H17. Genomics generating chaos means that it provides fodder needed to sustain theory builders.

H18. Luckily, classical biology has a theoretical framework ready-made (evolution).

Q3. The researchers correctly identified a classic, well-known pattern but, what to call it?

H19. Evolution can serve as a unifying, quantitative theory of oncology.

H20. Evolution touches essentially every oncology aspect (aetiology, pathogenesis, tumor progression, morphology, prognosis, treatment).

H21. Cancer theory should explicitly strive to build an evolutionary narrative that connect genetic/genomic alterations with phenotypic traits and interactions with tumor ecology.

H22. Evolutionary ecology is the core notion of oncological theory.

H23. Evolutionary theory applied to oncology and medicine produces not only a coherency to scientific medicine but also beautiful new mathematics.

9.13 DRIVING CANCER IMMUNOTHERAPY

In 2000's decade, an idea emerged in scientists, physicians, and medical experts minds. Instead of man-made chemicals to treat cancer, let people unleash bodies power on malignancy. Clift proposed H/Q/Ps on driving cancer immunotherapy.[28]

H1. Immuno-oncology: training the immune system to fight off cancer.

H2. Immuno-oncology tasks: (1) enhance molecular targeting of cancer cells; (2) report killing rate by specific immune agents; and (3) direct immune cells to tumor destruction.

Q1. How does the immune system target tumor cells?

Q2. How can an assay's two-step PCR be used to quantify clonal diversity of immune cells?

Q3. What was one able to do?

H3. The approach kind of changed the scale of what one was able to do.

H4. To monitor serial changes in thymus-dependent lymphocyte (T cell) clones before/during/after therapy.

Q4. What is happening in the tumor?

Q5. What is growing in the patient farther down the road?

H5. Stromal marker changes reflect tumor changes.

Q6. What is the route?

Q7. What is going to happen in terms of the research area?

P1. Leaky and radioactive labels, for example, ^{51}Cr.

P2. Assays that can only provide users with an endpoint for cell killing.

Q8. What is happening?

H6. Technology xCELLigence enables to sample broad conditions spectrum with simple workflow.

Q9. What are customer data telling them?

Q10. What drugs will work best for every patient?

H7. To segregate patients according to what drugs will work best for them.

9.14 THE MITOCHONDRION: CANCER'S CRYSTAL BALL

While the mitochondrion is well known as a source of cellular energy and metabolism, it also integrates cellular signaling to make cell fate decisions (whether a cell is to live or die by the mitochondrial pathway of apoptosis). Letai proposed Qs and As on the mitochondrion.[29]

Q1. How do cancer mitochondria differ from those of normal cells?

A1. Most cancer chemotherapeutics kill cancer cells via the pathway, so that it is essential to understand how cancer mitochondria differ from those of normal cells.

Q2. How ready is a cell to respond to pro-apoptotic signaling?

A2. His group invented a tool called BH3 profiling that measures how ready a cell is to respond to pro-apoptotic signaling.

They found that response to a wider variety of agents, for example, conventional chemotherapy, B-cell lymphoma type-2 (BCL-2) inhibitors, and targeted pathway inhibitors, could be predicted via BH3 profiling. It is becoming clear that the most important job of precision medicine, that of assigning the right drugs to the right patient, is poorly done by genomics alone. He proposed a H.

H1. Functional studies of tumor cells assaying mitochondrial response to therapeutics are essential in guiding drug assignment in cancer.

9.15 ONE ALGORITHM. MANY CURES. ZERO CANCER?

Pedro Domingos wrote the book *The Master Algorithm* (MA) and raised the questions.[30]

Q1. How will the quest for the ultimate learning machine remake our world?

Langston[31] interviewed Pedro Domingos on his book *The Master Algorithm*.

Q2. What is machine learning (ML) and how might a person encounter it in a typical day?

A2. ML is the automation of discovery. It is like scientific method on steroids: formulate Hs, etc.

Q3. Why is it important for someone who is not a computer scientist to understand ML principles?

A3. Learning algorithms (LAs) make a lot of decisions; if LAs are a black box, you have no control.

Q4. Why focus on what different tribes in ML research might contribute to curing cancer?

A4. Curing cancer is one important problem in the world and ML has a big part to play in solving it.

Q5. What makes cancer hard?

A5. It is not one disease but many; every patient's cancer is different and it mutates as it grows.

Q6. What does every approach bring to the table?

Q7. What is it missing?

Q8. What is the difference between algorithms that Netflix/Amazon use to recommend you products?

A8. Neflix directs you to obscure British TV shows from 70s. Amazon is based on how popular every product is.

Q9. Why is it important for consumers to be aware of these differences?

A9. Problem is that you wind up doing what companies want you to do, instead of what you want.

Q10. How do the learning algorithms work?

Q11. How did Obama's chief scientist use four simple questions to help win the 2012 election?

A11. He used ML to predict As to four Qs for every individual swing voter via all data about them he could get his hands on.

The questions were the following.

Q12. How likely is he to support Obama?

Q13. To show up at the polls?

Q14. To respond to the campaign's reminders to do so?

Q15. And to change his mind about the election based on a conversation about a specific issue?

She proposed additional questions and answers on his book.

Q16. Which voters to target the following day based on the results of the machine learning?

Q17. The result?

A17. Even though the race was close, Obama carried all the swing states but one and won election.

Q18. How is a ML expert more like a farmer than a factory worker?

A18. An ML expert grows programs from data like a farmer grows crops from nutrients.

Q19. What is the relationship between machine learning and artificial intelligence (AI)?

A19. AI goal is to get computers to do things that required human intelligence, for example, learn from experience; ML is an AI subfield.

Q20. Is computers that go awry something to worry about or are there other potential dangers?

A20. People worry that computers will get smart and take over the world, but problem is that they are stupid and they took over world.

Q21. What has ML enabled university scientists/researchers to do that would not be possible?

A21. ML is revolutionizing science by making it possible to understand more complex phenomena.

Q22. How do large social networks, with millions or billions of people, behave?

Q23. What is the Master Algorithm (MA) and how far are we from finding it?

A23. MA is a single algorithm capable of discovering all knowledge from data (e.g., human brain, evolution).

Q24. When will we find it?

A24. It is hard to predict because scientific progress is not linear.

Glaser interviewed Domingos proposing Q/Hs on one algorithm, many cures and zero cancer.[32]

Q25. Zero cancer?

H1. (Domingos, 1995). Cancer can be eliminated if people get serious about ML.

H2. (Domingos, 1995). A thinking machine will offer accurate predictions about every cancer.

H3. (Domingos, 1995). The ultimate output will consist of treatments and cures.

Q26. How is ML different than what is commonly known as AI?

A26. The ML is a subfield of AI. Learning is arguably the most important aspect of intelligence.

Q27. What is driving the current wave of progress in AI?

A27. ML is driving the current wave of progress in AI.

Q28. What do you view as biggest misconceptions about ML and flaws in ML-skeptics arguments?

A28. People think that ML is much more limited than it really is because for most, it is a black box.

Q29. What does it go on inside machine learning?

A29. A learning algorithm is, at some level, a miniature brain.

Q30. To realize a scenario in which an MA cures cancer, would all biology/pathology have to be translated into digital information?

A30. Absolutely, every LA worth its salt can deal with grey/fuzzy knowledge and big/noisy data.

Q31. How much digitalization of biology and pathology needs to occur?

Q32. Is this an achievable goal now or in the near future?

H4. While this approach was successful with many diseases, it is unlikely
to work for cancer.

Q33. How do cells work?

Q34. What would the MA that you propose look like and be capable of in
discovering a cure for cancer?

A34. It would provide a detailed model of how cells work, both healthy
and cancerous.

Q35. How would that work?

A35. We would be able to instantiate that model to every patient/cancer
and probe it with drugs.

Q36. Is it more appropriate to talk about an MA discovering cures for
cancer?

A36. Exactly. There is no single cure for cancer. The real cure is a ML
system.

Q37. Why is access to patient data and clinical outcomes so important?

A37. Because every cancer is different, there is something to learn from
every patient.

Q38. How does it contribute to inverse deduction, which you describe as
first step in curing cancer?

A38. Generalizing from which treatments worked for which cancers and
which did not, we predict.

Q39. Can you describe what program CanceRx to input a cancer's genome
and output a drug will look like?

A39. CanceRx could be as simple or complex as using a detailed model of
how cells work to test drugs in silico.

Q40. How do cells work to test candidate drugs in silico?

Q41. Can you tell us how far along it is in development?

A41. Rapid progress is being made across the full spectrum but there is
still a long way to go.

Q42. Can some of the same concepts and uses of ML above be applied to
vaccine development?

Q43. When an infectious agent,for example, Zika virus, emerges, do you envision that an MA could identify a vaccine?

A43. Microsoft used ML to develop an AIDS vaccine; AIDS virus is a tough adversary because it mutates quickly.

9.16 MASSIVE-SEQUENCING GENE CAPTURE AND APPLICATION IN CANCER GENETICS

Martínez González proposed H/Qs on massive-sequencing gene capture and cancer genetics.[33]

H1. Change from pregenomic research (cf. Fig. 9.1a) to postgenomic research (Fig. 9.1b).

Q1. How to find a mutation associated with a disease?

Q2. What role does a change in one base play in genetic variation?

Q3. What changes can one find?

Q4. In addition, how is nucleic-acids sequencing done?

Q5. Germinal mutations: How to find a mutation associated with a disease?

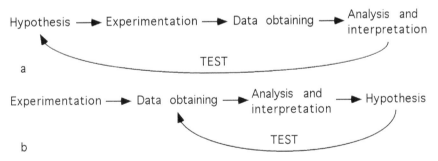

FIGURE 9.1 From (a) pregenomic research to (b) postgenomic research.

9.17 MICROBES AND CANCER: MORE THAN A HYPOTHESIS, LESS THAN A SOLUTION

Guerrero and Berlanga proposed Qs and Hs on microbes and cancer, hypothesis, and solution.[34]

Q1.   What could the emergence and cause of cancer be?

Q2.   Is cancer because of either external or internal reasons?

Q3.   Are there hereditary cancers?

Q4.   Are some persons more prone than others?

H1.   (Fibiger, 1926). Inflammatory H of cancer determined by worm *Spiroptera carcinoma*.

H2.   (Duran and Reynals, 1953). *Tumours Induced by Viruses and the Viral Theory of Cancer*.

H3.   (Koch, 1884). Postulates to secure that a certain microorganism is responsible for a disease.

Q5.   Are microbiota changes in patients with colon cancer disease precursors or a tumor-development consequence?

Q6.   What comes first, either the change in the microbiota or the development of cancer?

H4.   Resident bacteria were in some way promoting cancer.

Q7.   What do microbia do?

Q8.   How do microbia interact with each other and with people?

Q9.   How do microbia reply to environmental changes, for example, diet, lifestyle, etc.?

9.18 ENLISTING MYELOID EFFECTOR CELLS IN THE FIGHT VERSUS CANCER

Monoclonal antibodies (mAbs) became an important addition to chemo- and/or radiotherapy for the treatment of cancer. They can exert multiple effector functions that can lead to eradication of the tumor, for example, induction of apoptosis, growth inhibition, and initiation of complement-dependent lysis. The mAbs can recruit immune effector cells. Antibody-dependent cellular phagocytosis (ADCP) by macrophages is a prominent mechanism to remove tumor cells from the circulation after mAb therapy. Neutrophils potentially represent an important effector cell population that can eliminate tumor cells as they present potent cytotoxic ability, which is greatly enhanced in the presence of mAbs. Neutrophils play a role in regulating adaptive immunity. They may represent a large and potent effector cell population that can eliminate tumor cells, but may also induce antitumor immune responses. In order to investigate how mAbs mediate tumor cell elimination in vivo and to establish which effector cells are involved, van Egmond group used live cell and ex vivo imaging, and intravital microscopy.[35] Aiming to enhance myeloid cell recruitment and activity, mAb-based therapeutical strategies of cancer patients may be optimized.

9.19 INNOVATION IN THE ACCESS PROGRAMS IN ONCOLOGY

Grisolía, Lluch, and Camps organized Valencia Foundation of Advanced Studies Seminar Innovation in Access Programmes in Oncology. Berrocal proposed H/Q/A on immunotherapy and new therapeutic targets in cancer treatment.[36]

H1. Everybody generates cancerous cells that immune system bumps off but cancer develops mechanisms to evade immune system.

H2. There are four types of tumors that can be simplified in two: inflamed and noninflamed.

Q1. What does anticytotoxic T-lymphocyte antigen (CTLA)-4 bring forward?

A1. It brings forward more survival.

Q2. What is the following?

A2. Combination therapy is the future.

Q3. What problems must people solve?

A3. Absence of predictive factors and criteria for selection of patients.

A4. Absence of statistical methods of evaluating the effect with regard to other therapies.

H3. The used Bristol–Myers method assumes a kinetics that tends to extinction, which is false.

He concluded the following.

Conclusion (C)1. One lacks instruments to measure the efficacy and select the patients.

Massuti proposed Qs, Hs, and Ps on patients selection in oncology via biomarkers.[37]

Q4. Clinical decisions, future?

H4.　(Heisenberg, 1927). Uncertainty principle: $\Delta x \cdot \Delta p_x \geq h/4\pi$.

Q5.　Genomic information: targets, prognosis, …?

P1.　Selection tool is not definitive; tumor expression is not homogeneous depending on which point one make biopsy.

P2.　Immunogram: there are many points to be taken into account.

H5.　Medicines for healthy people: This is no innovation.

A round table discussed new-drugs approval in oncology and challenges from a regional perspective.

H6.　(Massuti). Quality requires centralization; dispersion implies a loss of quality.

H7.　(C. Camps). We live in a socialized medicine in a market economy.

H8.　(C. Camps). No market economy exists in pharma: no drug comes out halfway price of competence.

H9.　(C. Camps). The institutional pact is which, in the end, moves such theme.

H10.　(A. C. Cercós). Immunotherapy gives lesser secondary effects and lesser cost of these effects.

9.20　COMPETITION BETWEEN NORMAL AND TRANSFORMED EPITHELIAL CELLS

At the initial step of carcinogenesis, transformation occurs in a single cell within an epithelial sheet, and the transformed cells grow while being surrounded by normal epithelial cells. However, it was not clear what happens at the boundary between normal and transformed epithelial cells. Via newly established MDCK cell lines, Fujita group showed that when Ras- or Src-transformed cells are surrounded by normal epithelial cells, various signaling pathways are activated in the transformed cells and that they are usually eliminated from the apical surface of the epithelial monolayer.[38] The phenomena are not observed when transformed cells alone are present, suggesting that the presence of surrounding normal cells affects the signaling pathways and fate of transformed cells. Fujita group is analyzing the phenomena via mammalian cell culture and mouse model systems. Fujita presented their recent findings and proposed how the studies can elucidate molecular mechanisms regulating epithelial homeostasis and potentially lead to the establishment of novel types of cancer treatment: enhancing the ability of surrounding normal cells to fight versus cancer cells. He proposed H/Q/As.

H1.   Presence of surrounding normal cells affects signaling pathways and fate of transformed cells.

Q1.   How can studies elucidate molecular mechanisms regulating epithelial homeostasis and lead to establishment of novel types of cancer treatment?

Q2.   How is cell competition involved in tumorigenesis?

H2.   APC/RasV12 double-mutated cells invade the stromal tissue: APC mutation → Ras mutation [defect in epithelial defense against cancer (EDAC)!?] → invade the stromal tissue.

H3.   Cell competition is involved in the sequence of cancer.

Q3.   How is the metabolism regulated during apical extrusion?

Q4.   What is the relevance of these genes to humans, family-related cancer, etc.?

A4.   If a molecule is specific, it can be used as a marker.

Q5.   Your gene is a tumor suppressor, genetic tray?

Q6.   Is your glycolysis either aerobic or anerobic?

A6.   Our glycolysis is aerobic.

Q7.   In cancer cells, how long does it take with regard to normal cells?

Q8.   When the cell is extruded, does it die?

A8.   Ex vivo, yes.

Q9.   Is either the cell is extruded and dies, or the cell is mutated?

A9.   Yes, either.

Q10. In which state is oxidative phosphorylation impaired?

Q11. What does it happen to oxidative phosphorylation?

Q12. How did you study what happens to the extracellular matrix (ECM)?

Q13. How do these cells go to that point?

9.21   METASTATIC IMMUNE ESCAPE BY THE LOSS OF EXPRESSION OF INTERLEUKIN-33

Jefferies group described a new paradigm for understanding immune surveillance and escape in cancer.[39] Metastatic carcinomas express reduced levels of interleukin (IL)-33 and diminished levels of antigen processing machinery (APM), compared with syngeneic primary tumors. Complementation of IL-33 expression in metastatic tumors upregulates APM expression and functionality of major histocompatibility complex (MHC)-molecules, resulting in reduced tumor growth rates and a lower frequency of circulating tumor cells. Parallel studies in humans showed that low tumor expression of IL-33 is an immune biomarker associated with recurrent prostate and kidney renal clear cell carcinomas. The IL-33 presents a role in cancer immune-surveillance versus primary tumors, which is lost during the metastatic transition that actuates immune escape in cancer.

9.22 MOLECULAR-HETEROGENEITY CLINICAL IMPACT IN LARGE B-CELL LYMPHOMA

Large bone marrow-derived thymus-independent lymphocyte (B-cell) lymphoma is a heterogeneous group of lymphoid neoplasms that includes up to 16 different clinicopathological entities.[40] Advances in the molecular characterization of the diseases provide insights into the pathogenesis, evolution, and molecular markers of clinical outcome. Molecular characterization of lymphoma samples by different profiling methods allows for the identification of novel diagnostic markers that are used for further classification of the disease and contribute to the evolution of current diagnostic schemes. Deep sequencing methods are useful for the identification of novel genetic alterations that might explain particular phenotypes of the disease. The interaction between the novel molecular markers and clinical outcome is variable and a limited number of molecular markers show a real impact on patient prognosis or therapy selection. However, the availability of drugs that interfere with specific cellular pathways altered by somatic mutations in B-cell lymphoma is changing the paradigm of lymphoma treatment to a more molecularly oriented therapy that will probably boost the field of lymphoma molecular diagnosis in the near future.

9.23 THE SECRETS IN OUR CELLS

Cruse proposed the following questions and hypotheses on the secrets in our cells.[41]

Q1. How best to develop a successful cure for cancer?

Q2. Is it now possible that body's chemistry predicts one's future?

Q3. What has the future in store for us?

H1. A specific treatment benefits one.

H2. Metals–aggressive BC correlation occurs because as tumor grows, it needs additional support.

H3. Fact that metal concentration in patients' bodies assesses cancer severity develops treatments.

H4. Drugs lowering Fe content block the growth of some breast tumors.

H5. Drugs lowering Fe and binding Cu help doctors providing BC patients with treatments tailored to their body's only chemistry.

H6. Metal content is biomarker in BC, giving insight into individual disease characteristics/providing treatment.

H7. Personalized medicine: medical approach to healthcare looking to tailor treatment to every patient based on their only genetic code.

9.24 MUSCULOSKELETAL SARCOMAS: A MULTIDISCIPLINARY APPROACH

Amaya organized Valencia Foundation of Advanced Studies Day on musculoskeletal sarcomas with a multidiscilinary approach and F was proposed.

F1. (V. Boluda). Sarcomas are of low prevalence but deadly.

Amaya moderated a round table on basic approach to sarcomas.[42] Gomis raised Qs on metastasis and prediction/prevention of metastases specific sites.[43]

Q1. Can one predict and prevent specific sites of metastases?

Q2. How does one raise it?

Q3. How does one make it in the laboratory?

Q4. How does one translate it from mouse to human?

He presented cancer-progression/metastasis phases: (1) premalignant; (2) malignancy; (3) dissemination; (4) metastasis.

He proposed additional hypotheses, questions, and answer.

H1. Metastasis takes refuge in bones.

H2. Different organs select for distinct species of metastasis.

H3. Different BC subtypes select for distinct species of metastasis.

Q5. Can one predict tissue-specific metastasis?

Q6. Can one prevent tissue-specific metastasis?

Q7. How is it activated?

Q8. Why is it activated?

Q9. How can one make it in the laboratory?

H4. Oestrogen receptor (ER) status influences the choice of metastatic bone sarcome.

Q10. What does MAF gene do?

Q11. What does MAF know to make?

A11. It makes that cancer express in bone.

Q12. The administration of bisphosphonates functions in mice, what does one do with this?

Q13. It serves people, and what can one do to prevent metastasis in humans?

Q14. Is there an MAF–bisphosphonates successful interaction with the premenopause state?

H5. Bisphosphonates cause that metastasis does not adhere to bone and hang on other place.

Alonso proposed Qs, As, and H on human genetics and cancer programing.[44]

Q15. Is cancer programed?

Q16. How many genes are involved in cancer?

A16. There are 1571 (518 validated and 1053 candidates).

H6. (Knudson, 1971) Knudson model: two hits.

Q17. What is a single-nucleotide polymorphism (SNP)?

A17. The SNP is a variation in a single nucleotide that occurs at a specific position in the genome.

Q18. Where are people going?

A18. Precision medicine.

Q19. Is cancer programed?

A19. Yes, in a small percentage of patients.

Q20. What benefits does mutants identification in genes of predisposition to cancer present?

Q21. New genetic tests, want to know my future?

De Álava proposed Qs and A on molecular pathology of sarcomas.[45]

Q22. Why can one personalize therapy?

Q23. What is the actual impact of diagnosis on sarcoma treatment?

Q24. What is the impact of technical issues on the actual clinical performance of sarcoma treatment?

Q25. What is the indication to conduct molecular analyses?

A25. Sarcoma diagnosis is integrative.

A26. It is infrequent morphology and uncommon clinic.

Q26. What does one need to personalize therapy?

Additional questions and answers followed.

Q27. (Amaya). Are you close to assign to every case the adequate therapy?

A27. (Gomis). A good diagnostic is necessary; physicians search for the good biomarker.

Q28. (E. Bendala-Tufanisco). Is there a therapy that the suppressor genes lost in cancer recover?

A28. (De Álava). No, it would require editing 100% of the cancer cells.

A29. (Gomis). No, it would require editing the immune cells.

Q29. (Alonso). Would it make sense to design a drug against a nuclear transcription factor (NTF)?

A29. (Gomis). Yes, but it is difficult; an NTF cannot be inhibited.

Q30. (Medina). What is advantage of knowing the particular mutation as therapeutical approach?

A30. (Alonso). In a family, the physician can know what individuals are or not at risk.

A31. (De Álava). Some are essential for diagnosis; physician distinguishes them from similar others.

Q31. If a physician finds a variant in a gene in a cancer, how does he handle it?

A31. (Alonso). Physicians only inform mutations that are pathogenic.

Q32. How do physicians handle changing information?

A32. (Alonso). Physicians must revise it periodically.

Q33. Are physicians going ahead and endangering patients?

A33. (De Álava). Biomarkers require: (1) analytical validation with varied tools and (2) clinical validation.

Montalar moderated a round table on medical treatment of sarcomas.[46] Martín raised a Q on expectations in medical treatment of bone and soft-parts sarcomas.[47]

Q34. Are predictive biomarkers scarce in subtype sarcomas?

Cañete proposed Qs/A on peculiarities of the oncological treatment in paediatric age.[48]

Q35. What message does one want to give?

A35. Children are different.

Q36. Surgery (SUR) + radiotherapy (RT) + chemotherapy (CT)?

Q37. At what point are people in the Valencia Community (VC)?

Berlanga proposed Q/As on research in osteosarcoma biology and therapeutic relevance.[49]

Q38. What is osteosarcoma?

A38. It is the most frequent malign bone tumor in adolescents and young adults.

Q39. In addition, why to talk about osteosarcoma?

A39. It is the worst prognostic cancer: 15–20% of detectable metastasis.

Q40. Has osteoporosis bisphosphonate zoledronic acid (INN) antitumor effect in preclinical assays?

A40. It seems not.

Additional questions and answers followed.

Q41. What is adolescence?

A41. (Berlanga). World Health Organization (WHO): −18; VC: 15–; epidemiology: 15–19.

Q42. (Montalar). What is adjuvant therapy?

A42. (Martín). Adjuvant: in addition to primary therapy; neoadjuvant: before main treatment.

Q43. (Berlanga). Can a physician use adjuvant therapy after relapse?

A43. (Martín). Yes.

Q44. How to distinguish the tumor?

A44. (Berlanga). They are distinguished by analyses: (1) tumor and (2) out-of-tumor biopsy (blood).

Q45. (Montalar). Is adjuvant therapy more effective?

A45. (Martín). In this case, classic therapy is more effective.

Q46. (Díaz). In what histological subtypes is there mass reduction?

A46. (Martín). In neoadjuvant therapy, mass reduction <30%; in all, but peripheral tissue is least.

Q47. (Díaz). What is the role of adjuvant therapy?

Q48. (Díaz). How do you combine adjuvant therapy?

A48. (Martín). I would recommend RT or SUR + RT.

Q49. (Díaz). What is the population in which such an assay has been carried out?

A49. (Berlanga). Solve it with a randomized study?

Q50. Was it by the model or by the pharmaceutical (pharma) companies?

A50. (Berlanga). Both: the model is old, but the study was more academic than of the pharma.

Baixauli moderated a round table on oncology tumor surgery.[50] Vélez proposed Q/As on application of new technologies in sarcomas surgical planning.[51]

Q51. Pelvic sarcomas surgery, adjuvant therapy?

Q52. Can computer tools assist sarcomas surgery?

Q53. Margins in osteotomy images?

Q54. Is safe having only bone image?

Q55. (Baixauli). In this surgery, are physicians reaching the limit?

A55. (Vélez). It is one more tool; it always helps.

Q56. What is the limiting factor?

A56. (Vélez). It is time: before, with a Belgian company, 15 days; now, with a local one, 1 week.

Q57. Who does he draw that tumor?

A57. (Vélez). Between the radiologist and the surgeon.

Q58. (Baixauli). What does it happen if osteotomy image delays?

A58. (Vélez). The physician must bring forward adjuvant therapy.

Angulo proposed F on trends in giant-cells tumors (GCTs) treatment and Denosumab effect.[52]

F2. GCT is not sarcoma but benign tumor; the problem is frequent relapse.

He raised open questions on new trends in the treatment of GCTs.

Q59. What is the optimal length of the treatment?

Q60. Does it decrease the relapse rate?

Q61. Are there long-range adverse effects?

Q62. Must a physician modify the strategy?

Puertas raised a Q on extremity-saving surgery versus amputation and when amputating.[53]

Q63. When to amputate?

Gracia raised a question on evolution of the surgical treatment of pelvic sarcomas.[54]

Q64. Is resection of tumors involving the pelvic ring justified?

Mora and Edo raised Q on musculoskeletal tumors experience from patient viewpoint.[55]

Q65. (Edo). It is like when one washes his hands that do not become completely clean without soap, is not?

9.25 SOMATIC MUTATIONAL PROCESSES AND CANCER VULNERABILITIES

Somatic mutations are the driving force of cancer genome evolution. Somatic mutations rate results greatly variable across the genome because of variations in chromatin organization, DNA accessibility, and replication timing. However, other variables that may influence the mutation rate locally are unknown. López-Bigas group showed that DNA-bound proteins (e.g., transcription factors, histones) interfere with the nucleotide excision repair (NER) machinery, which increases DNA mutations rate at the protein binding sites.[56] The finding presents implications for people's understanding of mutational and DNA repair processes, and in the identification of cancer driver mutations. Given cancer evolutionary principles, one effective way to identify genomic elements involved in cancer is tracing the signals left by the positive selection of driver mutations across tumors. They identified 459 cancer genes with driver mutations analyzing 7000 tumor exomes from 28 different cancer types, and they searched for their targeted therapeutic opportunities. They analyzed hundreds of tumor whole-genomes to identify noncoding elements [e.g., promoters, enhances, 5' and 3' untranslated regions, micro (mi)ribonucleic acids (RNAs), long noncoding (lnc)RNAs], with cancer driver mutations. She proposed H and Qs.

H1. Cancer mutations give a lot of information on basic biology.

Q1. Do DNA-binding proteins influence the mutation rate?

Q2. Is mutation site in transcription-factor binding site (TFBS) also risen in other tumor sites?

Q3. Are there driver mutations (drivers) in noncoding regions?

She proposed four key questions (KQs).

Q4. KQ1. How many drivers has every tumor?

Q5. KQ2. How many are coding and noncoding?

Q6. KQ3. How many mutational and structural variants are there?

Q7. KQ4. How many therapeutic opportunities are there?

She rose an additional question.

Q8. How many patients could benefit from current and future targeted drugs versus cancer drivers?

She proposed additional questions and answer on a driver.

Q9. Is in a cancer gene?

A9. Not all mutations in driver genes are in cancer genes!

Q10. Is a driver mutation?

She proposed additional questions and answers.

Q11. Are there passenger mutation (passenger)–driver associations?

A11. Yes, some are synergistic, some others are at large distance.

Q12. Have you seen triggering by the driver?

A12. Yes, there are correlations.

Q13. Is there complexity?

A13. We saw every mutation and gene independently but, in the real world, complexity exists.

Q14. Many genomic effects reduce mutation rate?

A14. I never thought it: it is not easy to program.

Q15. Do they maintain a state of either many or few mutations?

A15. Mutations have information on what genes are more active/inactive (up/down) and open/close.

Q16. You see more mutations in the nucleosome, is it because chromatin is there?

A16. We have neither a map of damage nor a map of repair.

Q17. Can you predict if the damage affect?

A17. We can predict different kinds of damages; how a property changes needs a map of damage.

Q18. How does a property change?

Q19. What do you observe focusing on selection?

A19. On selection, we observe not much negative selection: for this, we need a mutation model.

9.26   INSIDE-OUT: INTRACELLULAR STRESS AND ORGANISMAL HOMEOSTASIS

In response to specific triggers, cancer cells die in a way that promotes a therapeutically-relevant tumor-specific immune response.[57] Such an immunogenic cell death relies on the activation of adaptive stress responses in dying cancer cells, which are required for the emission of immunostimulatory danger signals and consequent antitumor immunity. Many anticancer agents currently employed in the clinic are unable to initiate this process, calling for the development of combinatorial regimens to boost the immunogenicity of cancer cell death in patients. Adaptive stress responses in cancer cells constitute promising target to improve the immunological control of cancer in clinical scenarios.

9.27   pH HOMEOSTASIS AND CANCER

Cancer is a disease of mechanisms controlling cell growth and proliferation.[58] In the past, emphasis was on biochemical mechanisms of signal transduction involving growth factors and their receptors, protein kinases, G proteins, transcription factors, and chromatin modification systems.

Cell growth requires biophysical mechanisms involved in proton extrusion and, accordingly, tumors present higher intracellular (pH_i) and lower extracellular (pH_o) pH than normal tissues. High pH_i was considered a permissive factor without regulatory function. However, genetic manipulation of pH_i in mouse fibroblast with a yeast proton pumping ATPase resulted in tumorigenic transformation. In addition, low pH_o is required for metastasis. Two important things prevented consideration of the results by the scientific community: ignorance of mechanisms and lack of therapeutical consequences. The obstacles are in process of been overcome: mechanisms of pH_i and pH_o effects on cell growth and proliferation are being clarified and novel therapies are emerging. Low pH_o activates secreted lysosomal proteases promoting disruption of extracellular matrix and it can be counteracted by bicarbonate. High pH_i seems to activate TORC1 and G1 cyclins, and could be counteracted by weak organic acids.

9.28 CERN AND HADRON THERAPY: FROM PHYSICS TO MEDICAL APPLICATIONS

Dosanjh proposed Q/A/H on Conseil Européen pour la Recherche Nucléaire (CERN) and hadron therapy (HT).[59]

Q1. How to meet challenges of medical application?

Q2. Why is cancer important?

Q3. Tumor: why?

A3. Abnormal growth of cells; malignant: uncontrolled, can spread to cancer.

Q4. Treatment: how?

A4. Surgery, radiation (RT), and chemotherapy [CT via drugs (anticancer agents)].

H1. (Wilson, 1946). Protons can be used clinically.

Q5. What is a tissue?

Q6. How to achieve proton p^+ beam yielding maximum therapeutic effect?

Q7. The future: magnetic resonance image (MRI)-guided p^+ therapy (PT)?

Q8. What needs to be done?

Q9. How to tackle organ motion, real-time image and feedback?

Q10. (J. Bernabeu). Are two clinical-problems types: monitoring with a positron emission tomography (PET) to see effect; for moving bodies, monitoring acceleration via MRI?

A10. Use field to track MRI p^+/cyclotron for detection in real time; see PT in real-time/get feedback.

Q11. What is the difference between HT and X-rays?

A11. The function that HT can deliver is not possible with X-rays.

9.29 ORTHOXENOGRAFTS: A STRATEGY TO PERSONALIZE PATIENT'S TREATMENT

To personalize the tumor patient treatment in mice is a complex process that begins in the moment of tumor resection or after taking tumor biopsy. Several variables should be taken in consideration via the process in mice: (1) the site of tumor (subcutaneous vs. orthotopic) implantation; (2) tumor-taking rates; (3) synchronize tumor-time growth in patients and mice; (4) time needed for mice treatment response and window of time available for patient treatment; (5) extension of the genetic tumor characterization (driving mutations vs. exome sequence); (6) categorization versus histogenetic properties; (7) existence of solid evidences about the correlation among specific genetic tumor alterations and drug responses; (8) concordance among drug response for subcutaneous versus orthotopic engrafted tumors; (9) can one treat patients with the best drug response identified in mice? really applied in patients?; and (10) it is a feasible strategy to apply in large series of patients. In order to go in depth in the analysis of the variables, Villanueva group engrafted the same primary human tumors [mainly colorectal, epithelial ovarian cancer (EOC), and lung tumors] subcutaneous and/or orthotopically (orthoxenografts) in nude mice.[60] Their work indicates that although generation of orthoxenografts is a more complex and expensive process, they generated evidences that orthoxenografts should have important advantages to personalize patient's treatment.

9.30   INTERVIEW WITH MARIA BLANCO: WE LIVE SOME MORE BUT WITH DISEASES

Sapiña interviewed Blasco, proposing questions and As on living some more but with diseases.[61]

Q1. Why are you in favor of attacking aging as a whole rather than the diseases that derive?

A1. Aging, which is a molecular process, is the origin of many diseases that affect our society.

Q2. This involves prolonging the years of life, has your group prolonged them in mice?

A2. In humans, life has already been prolonged; people live some more but with diseases.

Q3. Did you obtain that such mice that live some more, did it without suffering from more cancer?

A3. Indeed, this was one of the important subjects via enzyme telomerase.

Q4. In addition, how did you obtain it?

A4. We obtained it by activating telomerase in a temporary way.

Q5. Will generated mice that presented telomers double their natural species live some more time?

A5. They live some more.

Q6. Will they suffer more from cancer?

A6. They suffer less from cancer.

Q7. Does it happen despite the fact that tumor cells use telomerase…?

A7. Tumor cells must activate telomerase in order to become a cancer.

Q8. Is it a temporary activation?

A8. We use for this a gene therapy, with vectors that do not integrate into genome and telomerase gene dilutes.

Q9. Can such therapies contribute to make people to lower their guard in prevention?

A9. I do not think so, people are advancing and conscience rises that they should follow healthy life habits.

Q10. Is there a limit of life for the human species?

A10. It seems yes.

Q11. Are people programed, evolutionarily speaking, to arrive to a certain age?

A11. There are two concepts: health span (40–50 years) and longevity (125 years).

Q12. What is importance of genetics and environment in telomerase production/telomers length?

A12. The genetic part is important but there is already an epigenetic, environmental aspect.

Q13. What is to know the length of our telomers for?

A13. It can be used for getting an idea, a kind of red light.

Q14. What can people do right now with such information?

A14. If a person has short telomers, he will probably have to change his life habits, pass revisions regularly, etc.

Q15. Mice with activated telomerase present greater survival after a heart attack, does it open door to therapy?

A16. Sure! It is a treatment for heart attack, aplastic anaemia, pulmonary fibrosis, etc.

Q17. In addition, is it effective?

A17. Yes it is; first, it will be to go to diseases that present no treatment and to see if it could serve.

Q18. Nowadays, there are many groups investigating in aging, why does this field generate so much interest?

A18. Because we know that diseases killing in developed countries (heart attack/degenerative/cancer) are related to aging.

Q19. How was it to work in Carol Greider's laboratory?

A19. When I was with her, telomerase gene was not isolated and it was not known if it were important for cancer/aging.

Q20. What memories have you of such laboratory?

A20. It was not known if telomerase activity would be important in rest of organisms because it was found in a rare organism.

Q21. What did it happen to you arriving to head of Oncology Investigations National Centre (CNIO)?

A21. It is a mistake that an excellence-oriented center be headed by something that not be a researcher.

Q22. What were the challenges?

A22. We have encouraged innovation.

Q23. How can one move closer the research that is made from centers like CNIO to society?

A23. The first thing that I did was to create a communication department.

Q24. You and journalist Salomone published book *To Die Young, at Age 140*, how was such experience?[62]

A24. I was suggested to write book alone and replied I could explain it to a journalist and write it together.

Q25. Does spreading force one to think a more accessible discourse, rethink aspects of his research?

A25. It is a reflection on what others and I do because not only my group but also many others' research appears.

9.31 ANALYZING CELL DEATH: THE PATH TO CELL VIABILITY

Journal *Gen. Eng. Biotechnol. News* raised Qs on analyzing cell death and path to cell viability.[63]

Q1. How diverse can the paths to cell death be?

Q2. How can the right tools make negotiating those paths both easy and enlightening?

Q3 Were the molecular pathways associated with cell death modalities active or passive?

Q4. Were the molecular pathways associated with cell death modalities oxidative or apoptotic?

Q5. (Green and Levine, 2014). To be or not to be?

Q6. (Green and Levine, 2014). How do selective autophagy and cell death govern cell fate?

Q7. How did researchers use intracellular combinatorial libraries to select antibodies protecting cells from death?

Q8. How was cell-death prevention accomplished?

Q9. Is phosphatidylserine (PS) an apoptotic marker or a neurone saviour?

Q10. How does PS receptor (PSR)-1 recognize and remove apoptotic cells?

Q11. Has human PSR the capacity to repair injured axons?

Q12. How does PSR-1 remove cells via apoptosis and necrosis?

Q13. How does the plague manage to evade the immune system?

Q14. What mechanisms does bacterium *Yersina pestis* share with other organisms?

Q15. How can bacteria subvert apoptotic cell death by directly destroying Fas ligand (FasL)?

Q16. How did Lathem et al. add plasminogen activator protease (Pla) to glass slides with fluorescently tagged proteins?

Q17. Did such protease show an affinity for a specific protein?

Q18. What proteins were affected?

ACKNOWLEDGMENTS

The authors thank support from Generalitat Valenciana (Project No. PROMETEO/2016/094) and Universidad Católica de Valencia *San Vicente Mártir* (Project Nos. UCV.PRO.17-18.AIV.03 and 2019-217-001).

KEYWORDS

- computational systems biology
- mathematical oncology
- association
- causation
- genomic complexity
- cancer genome

REFERENCES

1. Mukherjee, S. *The Emperor of All Maladies: A Biography of Cancer*; Scribner: New York, 2011.

2. Barillot, E.; Calzone, L.; Hupé, P.; Vert, J. P.; Zinovyev, A. *Computational Systems Biology of Cancer*; CRC: Boca Raton, FL, 2013.

3. Kuang, Y.; Nagy, J. D.; Eikenberry, S. E. *Introduction to Mathematical Oncology*; CRC: Boca Raton, FL, 2016.

4. Torrens, F; Castellano, G. Modelling of Complex Multicellular Systems: Tumour–immune Cells Competition. *Chem. Central J.* **2009,** *3* (I), 75–1-1.

5. Torrens, F.; Castellano, G. Information Theoretic Entropy for Molecular Classification: Oxadiazolamines as Potential Therapeutic Agents. *Curr. Comput.-Aided Drug Des.* **2013,** *9,* 241–253.

6. Torrens, F.; Castellano, G. Molecular Classification of 5-amino-2-aroylquinolines and 4-aroyl-6,7,8-trimethoxyquinolines as Highly Potent Tubulin Polymerization Inhibitors. *Int. J. Chemoinf. Chem. Eng.* **2013,** *3* (2), 1–26.

7. Estrela, J. M.; Mena, S.; Obrador, E.; Benlloch, M.; Castellano, G.; Salvador, R.; Dellinger, R. W. Polyphenolic Phytochemicals in Cancer Prevention and Therapy: Bioavailability *Versus* Bioefficacy. *J. Med. Chem.* **2017,** *60,* 9413–9436.

8. Torrens, F.; Castellano, G. Molecular Classification of Antitubulin Agents with Indole Ring Binding at Colchicine-binding Site. In *Molecular Insight of Drug Design*; Parikesit, A. A., Ed.; InTechOpen: Vienna, 2018, pp 47–67.

9. Torrens, F.; Castellano, G. Molecular Classification of 2-Phenylindole-3-carbaldehydes as Potential Antimitotic Agents in Human Breast Cancer Cells. In *Theoretical Models and Experimental Approaches in Physical Chemistry: Research Methodology and Practical Methods*; Haghi, A. K., Thomas, S., Praveen, K. M., Pai, A. R., Eds.; Apple Academic–CRC: Waretown, NJ, In press.

10. Torrens, F.; Castellano, G. AIDS Destroys Immune Defences: Hypothesis. *New Front. Chem.* **2014,** *23,* 11–20.

11. Torrens-Zaragozá, F.; Castellano-Estornell, G. Emergence, spread and uncontrolled Ebola outbreak. *Basic Clin. Pharmacol. Toxicol.* 2015, 117 (Suppl. 2) 38–38.

12. Torrens, F.; Castellano, G. 2014 Spread/Uncontrolled Ebola Outbreak. *New Front. Chem.* **2015,** *24,* 81–91.

13. Torrens, F.; Castellano, G. Ebola Virus Disease: Questions, Ideas, Hypotheses and Models. *Pharmaceuticals* **2016,** *9,* 14–6-6.

14. Hill, A. B. The Environment and Disease: Association or Causation? *Proc. R. Soc. Med.* 1965, 58(5), 295–300.

15. Hanahan, D.; Weinberg, R. A. The Hallmarks of Cancer. *Cell* **2000,** *100,* 57–70.

16. Hanahan, D.; Weinberg, R. A. The Hallmarks of Cancer: The Next Generation. *Cell* **2011,** *144,* 646–674.

17. Tian, F. Where NGS Leads. *Eur. Biopharm. Rev.* **2013,** *63,* 46–49.

18. Berger, M. F.; Lawrence, M. S.; Demichelis, F.; Drier, Y.; Cibulskis, K.; Sivachenko, A. Y.; Sboner, A.; Esgueva, R.; Pflueger, D.; Sougnez, C.; Onofrio, R.; Carter, S.L.; Park, K.; Habegger, L.; Ambrogio, L.; Fennell, T.; Parkin, M.; Saksena, G.; Voet, D.; Ramos, A. H.; Pugh, T. J.; Wilkinson, J.; Fisher, S.; Winckler, W.; Mahan, S.; Ardlie, K.; Baldwin, J.; Simons, J. W.; Kitabayashi, N.; MacDonald, T. Y.; Kantoff, P. W.; Chin, L.; Gabriel, S. B.; Gerstein, M. B.; Golub, T. R.; Meyerson, M.; Tewari, A.; Lander, E. S.; Getz, G.; Rubin, M. A.; Garraway, L. A. The Genomic Complexity of Primary Human Prostate Cancer. *Nature (London)* **2011,** *470,* 214–220.

19. The Cancer Genome Atlas Network. Comprehensive Molecular Portraits of Human Breast Tumours. *Nature (London)* **2012,** *490,* 61–70.

20. Forbes, S. A.; Bindal, N.; Bamford, S.; Cole, C.; Kok, C. Y.; Beare, D.; Jia, M.; Shepherd, R.; Leung, K.; Menzies, A.; Teague, J. W.; Campbell, P. J.; Stratton, M.

R.; Futreal, P. A. COSMIC: Mining Complete Cancer Genomes in the Catalogue of Somatic Mutations in Cancer. *Nucleic Acids Res.* **2011**, *39*, D945–D950.

21. Salas, M. D. personal communication.

22. Majumder, D.; Mukherjee, A. Multi-scale Modeling Approaches in Systems Biology Towards the Assessment of Cancer Treatment Dynamics: Adoption of Middle-out Rationalist Approach. *Adv. Cancer Res. Treat.* **2013**, *2013*, 587889-1-26.

23. Sherratt, J. A. Predictive Mathematical Modeling in Metastasis. In *Metastasis Research Protocols*; Brooks, S. A., Schumacher, U., Eds.; Methods in Molecular Medicine No. 57, Humana: Totowa. NJ, 2001, Vol. 1, pp 309–315.

24. Buetow, K. personal communication.

25. Lluch, A. personal communication.

26. Chowell, D.; Boddy, A. M.; Mallo, D.; Tollis, M.; Maley, C. C. When (Distant) Relatives Stay Too Long: Implications for Cancer Medicine. *Genome Biol.* **2016**, *17*, 34-1-3.

27. Calabuig Fariñas, S. Book of Abstracts, Encuentro de Investigadores en Cáncer: *Dando la Cara por la Sociedad*, Alcoi, Alacant, March 15, Spain, 2016, FISABIO, Alcoi, Alacant, Spain, 2016, O-1.

28. Clift, I. C. Driving Cancer Immunotherapy. *Genetic Eng. Biotechnol. News* **2016**, *2016* (Apr. 1), 1–35.

29. Letai, A. G. personal communication.

30. Domingos, P. *The Master Algorithm: How the Quest for the Ultimate Learning Machine Will Remake Our World*; Basic: New York, 2015.

31. Langston, J. A. Q & A with Pedro Domingos: Author of *The Master Algorithm. UW Today* 2015, September 17, 1–7.

32. Glaser, V. One Algorithm. Many Cures. Zero Cancer? *Genetic Eng. Biotechnol. News* **2016**, *36* (11), 30–31.

33. Martínez González, L. J. Personal Communication.

34. Guerrero, R.; Berlanga, M. Microbis i càncer: Més que una hipòtesi, menys que una solució. *Mètode* **2016**, *90*, 108–110.

35. Van Egmond, M. Personal Communication.

36. Berrocal, A. Book of Abstracts, Seminario Innovación en los Programas de Acceso en Oncología, València, Spain, October 8, 2016, Fundación Valenciana de Estudios Avanzados, València, Spain, 2016, O-1.

37. Massuti, B. Book of Abstracts, Seminario Innovación en los Programas de Acceso en Oncología, València, Spain, October 18, 2016, Fundación Valenciana de Estudios Avanzados, València, Spain, 2016, O-2.

38. Fujita, Y. Personal Communication.

39. Saranchova, I.; Han, J.; Huang, H.; Fenninger, F.; Choi, K. B.; Munro, L.; Pfeifer, C.; Welch, I.; Wyatt, A. W.; Fazli, L.; Gleave, M. E.; Jefferies, W. A. Discovery of a Metastatic Immune Escape Mechanism Initiated by the Loss of Expression of the Tumour Biomarker Interleukin-33. *Sci. Rep.* **2016**, *6*, 1-14.

40. Montes Moreno, S. Personal Communication.

41. Cruse, M. The Secrets in Our Cells. *Eur. Biopharm. Rev.* **2016**, *7*, 8–10.

42. Amaya, J. V. Book of Abstracts, Sarcomas Músculo Esqueléticos: Abordaje Multidiscilinar, València, Spain, January 26, 2017, Fundación Valenciana de Estudios Avanzados, València, Spain, 2017, O-1.

43. Gomis, R. Book of Abstracts, Sarcomas Músculo Esqueléticos: Abordaje Multidiscilinar, València, Spain, January 26, 2017, Fundación Valenciana de Estudios Avanzados, València, Spain, 2017., O-2.

44. Alonso, J. Book of Abstracts, Sarcomas Músculo Esqueléticos: Abordaje Multidiscilinar, València, Spain, January 26, 2017, Fundación Valenciana de Estudios Avanzados, València, Spain, 2017, O-3.

45. De Álava, E. Book of Abstracts, Sarcomas Músculo Esqueléticos: Abordaje Multidiscilinar, València, Spain, January 26, 2017, Fundación Valenciana de Estudios Avanzados, València, Spain, 2017, O-4.

46. Montalar, J. Book of Abstracts, Sarcomas Músculo Esqueléticos: Abordaje Multidiscilinar, València, Spain, January 26, 2017, Fundación Valenciana de Estudios Avanzados, València, Spain, 2017, O-5.

47. Martín, J. Book of Abstracts, Sarcomas Músculo Esqueléticos: Abordaje Multidiscilinar, València, Spain, January 26, 2017, Fundación Valenciana de Estudios Avanzados, València, Spain, 2017, O-6.

48. Cañete, A. Book of Abstracts, Sarcomas Músculo Esqueléticos: Abordaje Multidiscilinar, València, Spain, January 26, 2017, Fundación Valenciana de Estudios Avanzados, València, Spain, 2017, O-7.

49. Berlanga, P. Book of Abstracts, Sarcomas Músculo Esqueléticos: Abordaje Multidiscilinar, València, Spain, January 26, 2017, Fundación Valenciana de Estudios Avanzados, València, Spain, 2017, O-8.

50. Baixauli, F. Book of Abstracts, Sarcomas Músculo Esqueléticos: Abordaje Multidiscilinar, València, Spain, January 26, 2017, Fundación Valenciana de Estudios Avanzados, València, Spain, 2017, O-9.

51. Vélez, R. Book of Abstracts, Sarcomas Músculo Esqueléticos: Abordaje Multidiscilinar, València, Spain, January 26, 2017, Fundación Valenciana de Estudios Avanzados, València, Spain, 2017, O-10.

52. Angulo, M. Á. Book of Abstracts, Sarcomas Músculo Esqueléticos: Abordaje Multidiscilinar, València, Spain, January 26, 2017, Fundación Valenciana de Estudios Avanzados, València, Spain, 2017, O-11.

53. Puertas, P. Book of Abstracts, Sarcomas Músculo Esqueléticos: Abordaje Multidiscilinar, València, Spain, January 26, 2017, Fundación Valenciana de Estudios Avanzados, València, Spain, 2017, O-12.

54. Gracia, I. Book of Abstracts, Sarcomas Músculo Esqueléticos: Abordaje Multidiscilinar, València, Spain, January 26, 2017, Fundación Valenciana de Estudios Avanzados, València, Spain, 2017, O-13.

55. Mora, L.; Edo, B. Book of Abstracts, Sarcomas Músculo Esqueléticos: Abordaje Multidiscilinar, València, Spain, January 26, 2017, Fundación Valenciana de Estudios Avanzados, València, Spain, 2017, O-14.

56. López-Bigas, N. Personal Communication.

57. Galluzzi, L. Personal Communication.

58. Serrano, R. Personal Communication.

59. Dosanjh, M. Personal Communication.

60. Villanueva, A. Personal Communication.

61. Sapiña, L. Entrevista a Maria Blasco Directora del Centre Nacional d'Investigacions Oncològiques: *Vivim més, però amb Malalties. Mètode* **2017**, *93*, 12–17.

62. Blasco, M. A.; Salomone, M. G. *Morir Joven, a los 140: El Papel de los Telómeros en el Envejecimiento y la Historia de Cómo Trabajan los Científicos para Conseguir que Vivamos Más y Mejor*; Paidós: Barcelona, 2016.
63. Genetic Engineering and Biotechnology News. *Analyzing Cell Death: The Path to Cell Viability*; Mary Ann Liebert: New Rochelle, NY, 2017.

CHAPTER 10

PAIN AND PLEASURE

FRANCISCO TORRENS[1,*] and GLORIA CASTELLANO[2]

[1]*Institut Universitari de Ciència Molecular, Universitat de València, Edifici d'Instituts de Paterna, PO Box 22085, E-46071 València, Spain*

[2]*Departamento de Ciencias Experimentales y Matemáticas, Facultad de Veterinaria y Ciencias Experimentales, Universidad Catolica de Valencia San Vicente Mártir, Guillem de Castro-94, E-46001 València, Spain*

*Corresponding author. E-mail: torrens@uv.es

ABSTRACT

The aim of this work is to initiate a debate by suggesting a number of questions, which can arise when addressing subjects of pain and pleasure, in different fields, and providing, when possible, answers, hypotheses, and conclusions. Pain, *an unpleasant sensory and emotional experience associated with actual or potential tissue damage, or described in terms of such damage*, acts undoubtedly as an alarm system, and the integrity of the organism depends on its correct functioning. In basic research, people have models to study neuropathic pain and its pharmacological treatment, for example, surgical manipulation of spinal nerves, an approximation developed at the end of the 20th century, and the administration of chemical agents (antibiotics, antitumorals), developed at the beginning of 21st century and derived directly from clinical observation, as it is shown that the use of these drugs can cause peripheral neuropathies in patients. What does it make up the pain experience? Pain causes changes in N-acetylcysteine (NAC), giving rise to negative emotional states. The changes in NAC function promote consumption and motivation for high doses of opioids. On rising motivation and consumption of high doses, risk of scaling of doses and overdoses of opioids goes up. Opioids are safe if the guidelines indicated for pain treatment are followed.

10.1 INTRODUCTION

Pain, *an unpleasant sensory and emotional experience associated with actual or potential tissue damage, or described in terms of such damage*, acts undoubtedly as an alarm system, and the integrity of the organism depends on its correct functioning. In basic research, people have models to study neuropathic pain and its pharmacological treatment, for example, surgical manipulation of spinal nerves, an approximation developed at the end of the 20th century, and the administration of chemical agents (antibiotics, antitumorals), developed at the beginning of the 21st century and derived directly from clinical observation, as it is shown that the use of these drugs can cause peripheral neuropathies in patients.

The following question (Q) was raised on pain.

Q1. What does it make up the pain experience?

The aim of this work is to initiate a debate by suggesting a number of questions (Q), which can arise when addressing subjects of pain and pleasure, in different fields, and providing, when possible, answers (A), hypotheses (H), and conclusions (C).

10.2 ANIMAL MODELS OF NEUROPATHIC PAIN: ANTITUMORALS VERSUS PERIPHERAL

Pain, *an unpleasant sensory and emotional experience associated with actual or potential tissue damage, or described in terms of such damage*, acts undoubtedly as an alarm system, and the integrity of the organism depends on its correct functioning.[1] However, when pain chronifies, it loses its protective sense and turns into a clinical situation per se, causing the patient sensations of stress, anxiety, and even suffering. From the different types of chronic pain, one of the most problematic is neuropathic pain, which is defined as a pain originated by a nervous wound or dysfunction of the central nervous system. Although fortunately, it presents a low incidence with regard to the total of situations of chronic pain, it responds badly to the usual pharmacological treatment because it is not always possible to identify such wound or dysfunction, or because it is not possible to know its etiology. The design of animal models to study this type of pain was complicated, and only recently, it began to be studied in the laboratory. Although different experimental approximations exist to the study of neuropathic pain, two models of induction of peripheral neuropathy are interesting in rodents: the manipulation of peripheral nerves and administration of antitumorals. However, they are not the only ones in mimicking a situation of chronic pain and other possibilities exist, for example, the development of diabetic neuropathy by administration of streptozotocin to rats, or central wounds by administration of a chemical agent, although these models are lesser used, generally, because of the high incidence of physiological and behavioral modifications that they present, which considerably complicate nociception assessment in laboratory.

In basic research, people have models to study neuropathic pain and its pharmacological treatment, for example, surgical manipulation of spinal nerves, an approximation developed at the end of the 20th century, and the administration of chemical agents (antibiotics, antitumorals), developed at the beginning of 21st century and derived directly from clinical observation, as it is shown that the use of these drugs can cause peripheral neuropathies in patients. Although much of the most important advances in the knowledge of neuropathic pain was because of the use of the model of ligature of sciatic nerve (e.g., alterations in the concentration of neurotransmitters, activation of glial cells); nowadays, models based on the use of neoplastic agents are gaining acceptation among the scientific community in part by their correlation with human clinic. The comparison between the two models shows that the modifications that are produced in situations of nervous hurt are similar in the short and medium time (modifications in peripheral innervation, release of proinflammatory substances, etc.), that both are useful models because they allow improving people's knowledge of the origin and development of neuropathic pain, and the development of specific drugs, and that perhaps they could supply objective data to people for selecting the most adequate model to study different types of neuropathic pain.

10.3 PAIN AND PLEASURE, THE BRAIN CONTROLS PAIN AND PAIN IS AVOIDABLE

People do not think about pain frequently, unless it affects them directly, in which case they do not stop thinking about it.[2] To try to avoid or ease pain leads people to separate the hands from an alight stove or start costly research programs in search of more efficacious new analgesic drugs (i.e., more efficacious molecules). Pain is only information. It is like a flashing light that people see in the distance; the more rapidly it blinks, the stronger the pain is. However, pain is only a call to attention and has no utility, at least that the brain notices it. A simple paper leaf put between people and the flashing light will block the information and stop the pain, it does not matter the strong that this be. To know that pain has no power (not more than the one that people's mind want grant it) does not help people when they suffer, but it implies that drugs must be neither too much strong nor rough to stop it. They must be only intelligent. Many analgesics that people use are extracts processed from natural plants, their synthetic copies or chemical compounds close relatives of the natural plants on which they are based. Plants do not try to help people. Some active principles derived from plants are toxic and serve the vegetable like defense; in fact, it is which makes them efficacious as drugs. A substance that blocks nerves action can kill people, if it blocks the nerves that reach the heart, or can ease the pain derived from operation if it blocks the nerves that communicate the brain with the incision place. The field researchers that develop new drugs get excited when they find a new poisonous specimen of a plant, insect, frog, bacterium, or fungus.

Grisolía and de Andrés organized XI Yearly Day: Pain Is Avoidable.[3] De Andrés reported that brain controls abdominal pain and raised Qs on brain controlling abdominal pain.

Q1.   How does people's nervous system works?

Q2.   The pain expands and spreads, but what does it control it?

Q3.   What are the responses of people's brain?

Q4.   What brain centers do act at each moment?

A neuroregulation exists over local regulation. All works according to which regulates the brain, which receives information via the nervous system, and medical novelties go in this way: *neuromodulator techniques.*

When a patient arrives at a pain unit, it must be explained to him that his abdominal-pain problem is because of how his nervous system works, which could be sensitized by an appendicitis suffered in childhood, which generated a plastic change in his colon form, which change sensitized a zone of his nerves, which, in turn, send a signal to his spinal cord and from this to his brain, and the problem must be solved in his cord, putting him a control system. The patient must understand that regulation exists and, sometimes, the solution is not yet in the stomach but in the same brain. The pain expands/spreads and what controls it is the nervous system. It can be analyzed what people's brain responses are, what brain centers act at each moment and people know that in determined circumstances of intestinal pathologies, they have reflexes of their organs that travel information from trunk to brain, which act as symptomatology perpetuators (*visceral hypersensitivity*), that is, problems maintenance not by their viscera itself but enervation, that is, *it is the brain and nervous-character structures, which, in the end, perpetuate patient symptomatology.*

Díaz Insa organized a day and raised questions on migraines.[4]

Q5.   What does the new international classification of migraines change?

Q6.   What does such classification provide?

Q7.   Why has one got a headache?

Q8.   What must physicians know about migraines physiopathology to improve their patients?

Q9.   What clinical tests to carry out?

Q10. What patients with migraines to carry out to?

Q11. Are chronic migraines a paradigm of how pain is perpetuated?

Q12. What drugs do physicians use?

Q13. What patients to treat?

Q14. Is neurostimulation already a reality in migraines?

Q15. What is botulinum toxin (BTX) in migraines for?

Q16. How is BTX action in migraines?

Q17. Why is BTX useful in migraines?

Q18. Is calcitonin gene-related peptide a road to future treatments?

10.4   PAIN IN THE JOINTS

Grisolía and de Andrés organized Advanced Studies Valencia Foundation Day on Pain in Joints and raised a Q.[5]

Q1.   Why do joints hurt?

Martí Bonmatí proposed Qs and answers on joints, image, and radiological markers.[6]

Q2.   Why use magnetic resonance?

A2.   Because physicians see not only bone but also cartilage/sinovial structures and accurately.

Q3.   Why should radiography no longer be considered a surrogate outcome measure?

A3.   Because it is late and does not correlate with the pain.

Q4.   What do engineers do in a department of radiology?

A4.   They obtain the best images and extract metrics from them.

Q5.   How are these techniques done?

A5.   Physicians must agree with what techniques to do.

Milara Payá proposed Q/A on cartilage-degradation joint biomarkers, bone remodeling, and inflammation.[7]

Q6.   What is a biomarker?

A6.   Usually, it is a substance but it can also be an image.

He raised the following three questions on biomarkers.

Q7.   Can the clinician measure the biomarker?

Q8.   Does the biomarker add new information?

Q9.   Will the biomarker help the clinician to manage patients?
       He raised additional questions.

Q10. Inflammation in *osteoarthritis*?

Q11. How are biomarkers defined in *osteoarthritis*?

Q12. What recognized biomarkers are there in *osteoarthritis*?

Collado Cruz proposed Qs/A on joint chronic pain from nociception to neuropathy.[8]

Q13. Has central sensitizing been shown in patients with hip, knee, or hand osteoarthritis?

A13. Yes, in knee osteoarthritis.

Q14. In what extent can neuroplasticity fail in the face of changes produced in the nociceptive system by neurological lesion?

Q15. What extent do patients with osteoarthritis present neuropathic pain in?

Q16. What patients do them develop neuropathic pain in osteoarthritis?

Q17. Will neurojoint physiology change teaching and relationship with the traumatologist?

Alonso Pérez-Barquero proposed H, Qs, and As on exercise, sport, and joint pain.[9]

H1. The running myth: Is running good for the joints?

Q18. In addition, so what?

A18. Runners end in the consulting room.

Q19. Is running good?

A19. Transform vicious circles into virtuous ones: obesity ↔ sedentary lifestyle; running ↔ training.

Q20. Is running good for the joints?

A20. Running presents different effects on dissimilar persons; the A must be particularized.

Gomar Sancho raised Qs on knee, compartments, and kneecap role as stability element.[10]

Q21. Are people talking about chondromalacia (softening of the cartilage)?

Q22. What is the ethiopathogeny?

Q23. Bad sliding by an anatomical defect?

Q24. How to treat it?

Q25. Why does it hurt?

Q26. Is it because of the softening of the cartilage?

He proposed additional questions and answer on chondromalacia treatment.

Q27. Conservative treatment?

Q28. Surgical treatment?

Q29. Knee patches?

A29. They can atrophy the quadriceps.

Q30. Surgery?

Poveda Roda proposed Qs/As on craniofacial joint pain and temporomandibular joint.[11]

Q31. Why is the temporomandibular joint a singular one?

A31. Five reasons: it is placed at both sides of sagittal axis linked by an only bone; it has a meniscus; it depends on teeth; its pathology is usually treated by stomatologists; importantly, the great frequency with which it causes pain, especially chronic one.

Q32. Has stress (modern life, etc.) an effect on craniofacial joint pain?

A32. Yes, psychology (stress, depression, etc.) has an effect on craniofacial joint pain.

Asensio Samper raised a question on neuromodulation for the treatment of joint pain.[12]

Q33. How does radio frequency function?

Fabregat Cid raised Qs on chronic pain after joint surgery and reasons for the problem.[13]

Q34. Why does postoperative pain become chronic?

Q35. What does neuronal plasticity consist of?

Q36. In addition, in joint replacement?

Mínguez Martí proposed Q/As on pharmacological opportunities for joint pain beyond NSAID/opioids.[14]

Q37. Are bisphosphonates effective in the treatment of osteoarthritic pain?

Q38. Must the physician always put gastric protection?

A38. The A must be individualized (elder, with many drugs, especially corticoids, smoker, obese, etc.).

Q39. In patients with many drugs, to prescribe them opioids that other physicians remove, are they compatible?

A39. In opioids, the patient can have a cardiovascular disease treated by another physician.

Sanchis López raised the following questions on present and future intra-joint therapy.[15]

Q40. Will it be that it lacks lubrication?

Q41. Platelet-rich plasma, is it a new treatment for the rheumatologist?

10.5   ETHICS IN CRITIC/TERMINALLY-ILL-PATIENT CARE: DISEASE/SUFFERING MEANING

Zamora Marín made an analysis on the fundamentals of ethics and their association with medical services.[16] He compared abstract thought with natural sciences, noting that the postulates of the former should prompt good judgments of the physician at the patient's bedside. He stated a new health concept, including the ethical condition of the person. He stressed the importance of anthropological substantiation of ethics, partly via *Xavier Zubiri's* argumentation and personalism arguments, to analyze the main principles of ethics that should be taken into account in treating the seriously and terminally ill patients. He analyzed the concept of ethics based on values and virtues, trying to substantiate the expressed concepts and relate them to intensive and palliative cares. He discussed the concept of moral responsibility by *H. Jonas*, a phenomenon of humanistic control of science and their ethical implications observed in taking care of the seriously ill patient. He stated that delving more into Cuban cultural ethics and ethical formation of health professionals is a necessity, taking into consideration that they constitute in one way or another the ethical foundations of the Cuban nation. He explained the ethical bequest inherited from *Felix Varela* and *José de la Luz y Caballero*.

He addressed the problem of suffering in the sick person and defined it as the lack of well-being.[17] Not always can a moderate state of physical pain be considered suffering as such. The only way of eliminating suffering and turning it into joy and peace is to find the meaning. The patient who suffers should be capable of turning the meaning of his/her pain into an ally, aided by him/herself and by his/her physician. Beyond any doubt, this is the biggest challenge that a sick person may meet. Futile suffering is the logical explanation of why one true and fully suffers, the highest expression of which is the moral suffering that even when it disappears, it always leaves behind lasting impressions that continue to be the object of grief, which is the reason why the morale of every patient must be boosted, which is an action that is part of the art of curing as such, a task that all physicians are called upon to perform. The physicians, who are the ones that treat suffering people on a regular basis, should cultivate a culture of the meaning of suffering for their patients and them.

He analyzed differences between pain and suffering applying the concepts to the terminal patient.[18] The boundaries between both are essential to provide the help the patient needs the healthcare professional. The most feared symptom for the patient is suffering because of pain and end of life. However, the suffering will be mitigated, even eliminated, if he finds a sense for it to make him accept it.

10.6 PAIN IS AVOIDABLE: WE MUST HAVE AN OPERATION, SHOULD WE SUFFER PAIN?

De Andrés organized a Day on *Pain Is Avoidable* versus pain after surgery.[19] Pallardó Calatayud proposed Qs/As/Hs on pain perception/nociceptive routes in acute postoperative pain (APP).[20]

Q1. One must have an operation, should he suffer pain?

Q2. Pain perception and nociceptive routes in APP; are all people equal?

Q3. Are all people equal in the face of pain?

A3. No, because there are not two equal brains.

Q4. Why does not it hurt equally?

Q5. How can people deal with this?

Q6. Well, but where does it hurt exactly?

H1. Endogenous analgesia: areas of the cortex could block the periaqueductal gray.

Q7. What does it happen when one feels this painful feeling?

A7. Stress: hypothalamus–hypophysis axis stimulation, sympathetic nervous system activation, glucagon increment → reaction: pupil dilation, and so on.

Q8. What factors do affect the magnitude of postoperative pain (PP)?

A8. Patient idiosyncrasy, surgery, and so on.

H2. Pain perception is typical of every person being able to vary with conditions.

Q9. Pain modulation?

A9. It depends on a number of pain receptors in the zone, with depression (more), maniac (less), and so on.

Soriano Pastor proposed Qs, As, and H on adverse psychological effects of acute pain.[21]

Q10. Adverse psychological effects of acute pain, can one get ready for his surgery?

Q11. Why do women show greater pain?

Q12. *Anxiety* is the preoperative feature that associates the best with pain, why?

A12. *Depression* is associated with many modulator variables, why do not studies find preoperative depression–PP relation?

Q13. Why despite giving rational information to the patient he has an irrational idea?

A13. One should know how to give information (death thread → irrational); personal bias (Internet).

H3. Commitment to differential models.

Q14. A subject of how to give information to patients?

Goicoechea García proposed questions, A and Hs on pharmacogenomics and acute pain.[22]

Q15. How is genetics joining pharmacology?

A15. *Temporal summation*. It duplicates the shot speed: the normal is ×2 but someones ×800.

H4. *Pronociceptive* subjects: in upward routes, it goes faster; in downward routes, it goes slower.

H5. *Antinociceptive* subjects: in upward routes, it goes slower; in downward routes, it goes faster.

Q16. Epigenetics: Why is not your deoxyribonucleic acid your destiny?

H6. *Pharmacokinetics*: What organism does to the drug (transport, metabolism)?

H7. *Pharmacodynamics*: What drug does to the organism (receptors)?

Q17. (Hajj, 2015) Genotyping test with clinical factors: better management of APP?

Q18. Preventive analgesia, myth, or reality?

Monsalve Dolz proposed Hs and Q on psychological evaluation/ interventions in APP.[23]

H8. Model of cognitive–behavioral intervention: cognitive–behavioral therapy (CBT).

H9. CBT phases: psychodeductive, acquisition of skills; application of acquired skills.

H10. *Adherence*: if the patient knows antidepressants are administered to him for analgesia, less pain.

H11. *Information* requires to give and receive it.

H12. Some patients do not want to receive information; some others want: it is not *to give a note*.

H13. *Attention* is the most studied psychological variable (*anxiety* is the most studied medical one).

H14. *Attention* techniques: distraction (children/adolescents), guided imagination, visualization, musicotherapy, hypnosis (Ramón y Cajal, adults not APP).

Q19. In addition, with regard to the prevention of APP?

Sánchez Pérez proposed Q/A/H on drug administration routes/which are useful in APP.[24]

Q20. What drug administration routes are useful in the management of APP?

H15. (B. Obama). Precision medicine initiative, when can I sign up?

Q21. When can I sign up?

Q22. (Caba, 2010) Should epidural analgesia still be a routine technique in pain units?

Q23. What is sufentanil?

A23. It is a pure agonist of μ-opioid receptors.

Q24. Why sublingual route for sufentanil?

A24. Highly vascularized region [it delays 6.2 min, much lesser than intravenous (iv) morphine].

H16. The secondary effect of slow iv morphine is that patients do not note it and take overdoses.

H17. *Helsinki model* of acute pain.

Rodríguez Gimillo proposed Qs/As on patient-controlled/regional/local analgesia.[25]

Q25. Surgery = pain?

A25. Surgery $\neq$ pain.

Q26. When does it hurt?

A26. Operating theatre (anesthesia); postoperative: hospital (hospitable surgery), home (national health clinic surgery).

Q27. Regional analgesia: peripheral nerve blocking, how do people reach the nerve?

10.7   GRIEF OF PAIN: PROCESSING PHYSICAL PAIN/ ANALGESIA/EMOTIONAL COMPONENT

Hipólito proposed Qs/Hs/As on pain grief, processing, analgesia, and its emotional component (Hipólito, L., personal communication).

Q1.   How is painful sensation transmitted?

H1.   There are two types of pain: acute and chronic.

H2.   Factors: inflammatory (release of molecules causing inflammation), neuropathic (nerve wound).

Q2.   What drugs can people use?

A2.   Opioids (in brain), others (in spinal cord dorsal horn).

Q3.   How to avoid undesired effects?

A3.   Reduce dose, combine two drugs in subeffective doses (*synergism*), administer drugs locally.

H3.   Antiglutamatergic drugs, for example, ketamine, but they give important adverse effects.

H4.   Antiglutamatergic subeffective dose + other locally injected drug that opens K^+ channels.

Q4.   What drugs can people use?

A4.   Opioids, but they have two functions: analgesic and regulate reinforcements.

H5.   Opioids are the basis of both analgesia and addiction.

H6.   Circuit of reinforcement and motivation.

Q5.   Does pain affect mesocorticolimbic-system motivation areas, causing depression or addiction?

H7.   Animals with pain must receive higher doses to have the motivation, which has consequences.

H8.   Rats want higher doses of heroin while smaller doses are analgesic enough.

H9.   This is a danger for human patients that want higher doses while smaller are analgesic enough.

H10.   Human patients say that lower doses do not produce a high while higher doses do.

H11.   There is a drug that also activates μ-opioid receptors of reinforcement: ethanol.

Q6.   Does drinking help one to fight versus the grief of pain?

H12. Model of relapse: alcohol deprivation effect.

H13. Effect of relapsing into ethanol.

H14. In rats that do not relapse into ethanol, the pain increases the risk of relapsing into ethanol.

Q7.   A question of sex?

A7.   Animal experimentation: before, males (females affected by menstrual cycle); now, males + females.

Q8.   Are females more sensitive to relapse into ethanol?

10.8   EMOTIONAL DIMENSION OF PAIN: IMPACT ON THE ADDICTION TO OPIOIDS

Hipólito raised questions on emotional dimension of pain and impact on opioids addiction.

Q1.   When does pain persist?

Q2.   (Isabel Fariñas) Are people more inclined to be an addict if they suffer from pain?

      She proposed the following initial questions and answer on the emotional dimension of pain.

Q3.   Does a relationship exist between pain and addiction?

Q4.   Are patients with pain more vulnerable to suffer for an addiction?

Q5.   Are overdoses related to the treatment of pain or the hedonic effects of opioids?

Q6.   What is the cause of that the scaling of deaths by opioids in the United States does not occur in Europe?

A6.   Treatments that in Europe are not made with opioids, in the United States, they are easily treated with opioids.

      She proposed additional questions and answer on the emotional dimension of pain.

Q7.   How does pain function?

A7.   (International Association for the Study of Pain, IASP, 1994) Pain is an unpleasant *sensory* and *emotional* experience associated with actual or potential tissue damage.

Q8.   What medicines can people use?

Q9.   Pain presents an emotional dimension, addiction?
      She presented the following conclusions (Cs) on the emotional
      dimension of pain.

C1.   Pain causes changes in NAC, giving rise to negative emotional
      states.

C2.   The changes in NAC function promote consumption and motivation
      for high doses of opioids.

C3.   On rising motivation and consumption of high doses, risk of scaling
      of doses and overdoses of opioids goes up.

C4.   Opioids are safe if the guidelines indicated for pain treatment are
      followed.

10.9   STORIES OF MOTHER AND DAUGHTER CELLS: HOW TO ORGANIZE PEOPLE'S BRIAN

Gil Sanz raised questions (Q) on mother/daughter cells stories, and how to
organize our brain (Gil Sanz, C., personal communication).

Q1.   How to organize people's brain?

Q2.   How is gotten to generate cell diversity from a population of similar
      neural stem cells (NSCs)?

Q3.   How alterations in NSCs diversification are connected with neural
      disorders?

10.10   FINAL REMARKS

From the present results and discussion, the following final remarks can
be drawn.

1.   Pain causes changes in NAC, giving rise to negative emotional
     states.

2.   The changes in NAC function promote consumption and motiva-
     tion for high doses of opioids.

3.   On rising motivation and consumption of high doses, risk of
     scaling of doses, and overdoses of opioids goes up.

4. Opioids are safe if the guidelines indicated for pain treatment are followed.

ACKNOWLEDGMENTS

The authors thank the support from Generalitat Valenciana (Project No. PROMETEO/2016/094) and Universidad Católica de Valencia *San Vicente Mártir* (Project Nos. UCV.PRO.17-18.AIV.03 and 2019-217-001).

KEYWORDS

- **neuropathic pain**
- **antitumoral**
- **ligature of the sciatic nerve**
- **cannabinoid**
- **rat**
- **pain experience**

REFERENCES

1. Goicoechea, C. Modelos Animales de Dolor Neuropático: Dolor Inducido por Antitumorales frente a Manipulación de Nervios Periféricos. In *Aportaciones de los Estudios Funcionales a la Investigación Farmacológica Básica*; Badia, A., Ed.; Monografías Dr. Antonio Esteve No. 34, Fundación Dr. Antonio Esteve: Barcelona, 2008; pp 113–122.
2. Gray, T. *Molecules: The Elements and the Architecture of Everything*; Black Dog & Leventhal: New York, 2014.
3. Grisolía, S.; de Andrés, J. In *Book of Abstracts, XI Jornada Anual: El Dolor Es Evitable. El Sistema Gastrointestinal*, València, Spain, 2015; Fundación Valenciana de Estudios Avanzados: València, Spain, 2015; p O-1.
4. Díaz Insa, S. In *Book of Abstracts, Jornada Sobre Cefaleas y Migrañas*, València, Spain, 2016; Fundación Valenciana de Estudios Avanzados: València, Spain, 2016.
5. Grisolía, S.; de Andrés, J. In *Book of Abstracts, XII Jornada Anual el Dolor Es Evitable: Dolor en las Articulaciones*, València, Spain, November 10, 2016; Fundación Valenciana de Estudios Avanzados: València, Spain, 2016; p O-1.

6. Martí Bonmatí, L. In *Book of Abstracts, XII Jornada Anual el Dolor Es Evitable: Dolor en las Articulaciones*, València, Spain, November 10, 2016; Fundación Valenciana de Estudios Avanzados: València, Spain, 2016; p O-2.

7. Milara Payá, J. In *Book of Abstracts, XII Jornada Anual el Dolor Es Evitable: Dolor en las Articulaciones*, València, Spain, November 10, 2016; Fundación Valenciana de Estudios Avanzados: València, Spain, 2016; p O-3.

8. Collado Cruz, A. In *Book of Abstracts, XII Jornada Anual el Dolor Es Evitable: Dolor en las Articulaciones*, València, Spain, November 10, 2016; Fundación Valenciana de Estudios Avanzados: València, Spain, 2016; p O-4.

9. Alonso Pérez-Barquero, J. In *Book of Abstracts, XII Jornada Anual el Dolor Es Evitable: Dolor en las Articulaciones*, València, Spain, November 10, 2016; Fundación Valenciana de Estudios Avanzados: València, Spain, 2016; p O-5.

10. Gomar Sancho, F. In *Book of Abstracts, XII Jornada Anual el Dolor Es Evitable: Dolor en las Articulaciones*, València, Spain, November 10, 2016; Fundación Valenciana de Estudios Avanzados: València, Spain, 2016; p O-6.

11. Poveda Roda, R. In *Book of Abstracts, XII Jornada Anual el Dolor Es Evitable: Dolor en las Articulaciones*, València, Spain, November 10, 2016; Fundación Valenciana de Estudios Avanzados: València, Spain, 2016; p O-7.

12. Asensio Samper, J. M. In *Book of Abstracts, XII Jornada Anual el Dolor Es Evitable: Dolor en las Articulaciones*, València, Spain, November 10, 2016; Fundación Valenciana de Estudios Avanzados: València, Spain, 2016; p O-8.

13. Fabregat Cid, G. In *Book of Abstracts, XII Jornada Anual el Dolor Es Evitable: Dolor en las Articulaciones*, València, Spain, November 10, 2016; Fundación Valenciana de Estudios Avanzados: València, Spain, 2016; p O-9.

14. Mínguez Martí, A. In *Book of Abstracts, XII Jornada Anual el Dolor Es Evitable: Dolor en las Articulaciones*, València, Spain, November 10, 2016; Fundación Valenciana de Estudios Avanzados: València, Spain, 2016; p O-10.

15. Sanchis López, N. In *Book of Abstracts, XII Jornada Anual el Dolor Es Evitable: Dolor en las Articulaciones*, València, Spain, November 10, 2016; Fundación Valenciana de Estudios Avanzados: València, Spain, 2016; p O-11.

16. Zamora Marín, R. Ética en el cuidado del paciente grave y terminal. *Rev. Cubana Salud Públ.* **2006,** *32* (4), 1–7.

17. Zamora Marín, R. La enfermedad y el sentido del sufrimiento. *Rev. Cubana Salud Públ.* **2009,** *35* (1), 1–5.

18. Zamora-Marín, R. Algunas consideraciones sobre la enfermedad y el sentido del sufrimiento. *Therapeía* **2016,** *2016* (8), 121–125.

19. De Andrés, J., Ed., In *Book of Abstracts, XIII Jornada Anual: El Dolor Es Evitable,* València, Spain, November 16, 2017; Fundación Valenciana de Estudios Avanzados: València, Spain, 2017.

20. Pallardó Calatayud, F. V. In *Book of Abstracts, XIII Jornada Anual: El Dolor Es Evitable,* València, Spain, November 16, 2017; Fundación Valenciana de Estudios Avanzados: València, Spain, 2017; p O-1.

21. Soriano Pastor, J. In *Book of Abstracts, XIII Jornada Anual: El Dolor Es Evitable,* València, Spain, November 16, 2017; Fundación Valenciana de Estudios Avanzados: València, Spain, 2017; p O-2.

22. Goicoechea García, C. In *Book of Abstracts, XIII Jornada Anual: El Dolor Es Evitable*, València, Spain, November 16, 2017; Fundación Valenciana de Estudios Avanzados: València, Spain, 2017; p O-3.

23. Monsalve Dolz, V. In *Book of Abstracts, XIII Jornada Anual: El Dolor Es Evitable*, València, Spain, November 16, 2017; Fundación Valenciana de Estudios Avanzados: València, Spain, 2017; p O-9.

24. Sánchez Pérez, C.A. In *Book of Abstracts, XIII Jornada Anual: El Dolor Es Evitable*, València, Spain, November 16, 2017; Fundación Valenciana de Estudios Avanzados: València, Spain, 2017; p O-10.

25. Rodríguez Gimillo, P. In *Book of Abstracts, XIII Jornada Anual: El Dolor Es Evitable*, València, Spain, November 16, 2017; Fundación Valenciana de Estudios Avanzados: València, Spain, 2017; p O-11.

CHAPTER 11

LIPIDIC NANOPARTICLES: A PLATFORM FOR ADVANCEMENT IN DRUG DELIVERY SYSTEMS

PANKAJ DANGRE[1,*], KAUSHALENDRA CHATURVEDI[2], and PRASAD POFALI[3]

[1]R.C. Patel Institute of Pharmaceutical Education and Research, Shirpur, India

[2]Lachman Institute for Pharmaceutical Analysis Laboratory, Long Island University, New York, USA

[3]Shobhaben Pratapbhai Patel School of Pharmacy and Technology Management, SVKM's NIMMS, Mumbai, India

*Corresponding author. E-mail: pankaj_dangre@rediffmail.com

ABSTRACT

Nano-sized drug deliveries have emerged since last few decades with the objective of controlled and targeted delivery of drugs. Lipidic nanoparticles (LNs) witnesses promising alternatives for various colloidal drug delivery systems viz., nanoemulsion, liposomes and polymeric nanoparticles. The biodegradable and biocompatible nature of LNs proves itself favorable among other polymeric nanoparticles. Furthermore, LNs ensure better therapeutics by modifying drug release kinetics, bio-distribution, and greater uptake in tissues. In spite of potential ability of LNs for drug delivery their manufacturing still, pose a challenge. The present chapter presents insight on LNs including basic component, manufacturing techniques, characterization parameters, and applications toward drug delivery.

11.1   INTRODUCTION

In recent years, a large number of discovered drug molecules have low or no aqueous solubility. According to the United States of Food and Drug Administration, these drugs are classified as Biopharmaceutical Classification System (BCS) class II and IV.[1] Limited solubility of the drug in water results in poor bioavailability, and therefore, aqueous solubility of drug plays a major role in the success of drug delivery. It has become more and more evident that developing a new drug product alone is not sufficient to ensure progress in drug therapy. Creative and novel formulation efforts are needed to develop a drug product from BCS class II and IV active pharmaceutical ingredients (APIs) that have acceptable pharmacokinetic properties.[2] The best strategy to overcome these problems involves the formulation of a suitable drug carrier system. Based on the current approach toward the formulation development it can be clearly noted that the in-vivo fate of drug molecule is no longer determined by the properties of drug only, but by the carrier system, which would control or release of the active drug according to the specific needs of the therapy. Size of the carrier depends on the desired route of administration, which can be range from nanometers colloidal system to micrometer range as microparticles. There are multiple approaches to formulate a drug delivery system for BCS class II and IV drugs, which can lead to acceptable pharmacokinetic properties. One such approach is increasing the particle surface area by milling or homogenization, and nanoparticle formulation.[3,4] In recent years, there have been many advances related to lipidic nanoparticles (LNs) formulation system. It has several advantages, such as nonorganic solvent system, biodegradable nature, stability, cost-effective production, no biotoxicity, and high drug loading (DL). Above listed properties of LNs make it the preferred choice of formulation type for BCS class II and IV drugs. Many biocompatible lipids are in the solid state at room temperature and are inexpensive. Nano and microparticles made out of these lipids and suspended in water offer an excellent option for formulating biologics as well as small molecules. However, the development of LN is a complex process, and hence appropriate analytical techniques are needed for solid state and physical characterization of lipid nanoparticles. It is important to use analytical techniques to characterize LN formulation at particulate as well as molecular level.[5,6]

11.2 TYPES

11.2.1 SOLID LIPID NANOPARTICLES

Solid lipid nanoparticles (SLNs) were investigated initially in 1991 as alternative drug carrier approach to emulsions, polymeric nanoparticles, and liposomes.[7] SLNs are prepared using a solid lipid or a blend of solid lipids where the lipids stand in solid state at normal room temperature or body temperature. SLNs consist of solid lipids from 0.1% (w/w) to 30% (w/w) preferably dispersed in aqueous medium and stabilized using a suitable surfactant (0.5–5% w/w). The mean particle size of SLNs varies from 40 to 1000 nm.[8]

11.2.2 NANOSTRUCTURED LIPID CARRIERS

Nanostructured lipid carriers (NCLs) are second-generation lipid carriers produced using a blend of solid lipids and liquid lipids (oils). NCLs of desired sizes can be obtained by using suitable ratios of solid lipids and liquid lipids. The presence of liquid lipids (oils) in the blend lowers the melting points of solid lipids at room temperature; however, the overall mixture remains solid at body temperature. The solid lipids content in NCLs varies up to 95%.[9] NCLs shows high drug-loading capacity and avoid drug precipitation during storage.[10]

11.3 KEY FEATURES

LNs are considered promising colloidal carrier system for improving the oral bioavailability of poorly water-soluble drugs. The lipids employed in the fabrication of LNs are biocompatible, nontoxic, and generally regarded as safe (GRAS).[11] Moreover, its ability to improve the stability of drug, high DL, control or targeted drug release, as well as ease of manufacturing and scale up make it a versatile drug delivery system.[10,12]

11.4 FORMULATION ASPECTS

Formulation development plays a vital role while designing lipid nano-formulations irrespective of lipid, SLNs, and other nanotechnologies. The selection of lipid, surfactant, and cosurfactant for the best formulation, which renders reproducible results and characteristics governed by certain factors, is discussed below.

11.4.1 LIPIDS

Lipids play an immense role while designing the lipid formulation and lipid nanotechnologies and attracted the attention during recent years[10,13] and various parameters like chemical structure, ionic composition, and charges of the selected lipids affected both the microemulsion formation and the physical characteristics of lipid nanoparticles. Characteristic lipids with reference to lipid nanoparticles are referred in the following table:

Type of lipids	References
Acidan	[13–15]
Behenic acid	[16–18]
Butanol	[19,20]
Butyric acid	[21–23]
Cetyl palmitate	[24–26]
Decanoic acid	[27–29]
Dioctyl sodium sulfosuccinate	[30–32]
Egg lecithin	[33–35]
Hydrogenated coglycerides	[36–38]
Monooctylphosphoric acid sodium	[28,39,40]
Palmitic acid	[41,42]
Phosphatidylcholine	[43–46]
Stearic acid	[14,47]
Sodium cholate	[43,48]
Sodium glycocholate	[49,50]
Tricaprin	[51,52]
Trilaurin	[50,53]
Trimyristin	[16,54]
Tripalmitin	[45,55]
Tristearin	[51,55]
Taurocholic acid sodium salt	[56,57]
Taurodeoxycholic acid	[58,59]
Witepsol	[60,61]

11.4.2 SURFACTANTS

Surfactants impart essential characteristics to the LNs and, therefore, must be incorporated in suitable concentrations. It is also having the ability to incorporate more amount of drug. However, the nature and type of surfactant affect the particle size of LNs. The following surfactants are commonly employed in the fabrication of LNs.

Name of surfactants	References
Cremophor RH40	[62,63]
Polaxamer	[64,65]
Polysorbate 20 and 80	[66,67]
Polyvinyl alcohol	[68,69]
Sodium cholate	[43,70]
Soybean lecithin	[4,71]
Soybean phosphotadylcholine	[43,72]
Tyloxapol	[50]
Pluronic F68	[14,60]
Mono and dioctyl sodium sulfosuccinate	[73,74]
Gleceryl monostearate	[58,75]
Glyceryl behenate	[16]
Glyceryl palmitostearates	[55,76]
Poloxamine	[64,77]

11.4.3 COSURFACTANTS

Cosurfactants are adjuvant to the surfactant and needed to reduce the amount of surfactant in the formulation. Cosurfactants have the ability to dissolve more amount of drug as well as it stabilizes the charges on the surface of the particles.

Name of cosurfactants	References
Egg lecithin phosphotadylcholine	[78]
1-Butanol	[79]
1-Propanol	[79]
n-Pentanol	[79]
Ethanol	[79]

Name of cosurfactants	References
Sodium taurodeoxycholate (bile salt)	[80]
Pluronic F68	[80]
Tween 60 and 80	[80]
Amino acids (tyrosine, tryptophan, and phenylalanine)	[80]

11.5   METHOD OF PREPARATION

11.5.1   HIGH-PRESSURE HOMOGENIZATION

Lipid nanoparticles are particles made from lipids with average particle size ranges from 50 to 1000 nm. The microemulsion for parenteral nutrition can be produced by replacing the oil phase of the emulsion droplets by a lipid. The SLNs can be stabilized by surfactants and or polymers; however, emulsions for parenteral nutrition are stabilized generally with lecithin. Distinct ranges of lipopolymeric nanoparticles can be produced by high-pressure homogenization just like parenteral oil in water emulsions. The high-pressure homogenization has also been attributed as ultrahigh pressure where the pressure up to 350 MPa can be achieved to produce nanosize droplets of nanoemulsion.[81,82] High-pressure homogenization has been divided into hot and cold homogenization, which can be used for manufacturing the lipid nanoparticles.

11.5.1.1   HOT HOMOGENIZATION

When homogenization is carried out at a temperature above the melting point of lipid, it is known as hot homogenization. The device called Ultra-Turrax (High Shear Mixing Device) is quite more popular and can be used for synthesizing the pre-emulsion. More often, higher temperature results into lower particle size due to viscosity decrease of inner phase.[83] The product obtained primarily in case of hot homogenization is nanoemulsion due to lipid's liquid state. The oral bioavailability of SLNs of olmesartan medoxomil using hot homogenization method was found to be enhanced with lipid glyceryl monostearate and surfactant poloxamer for treatment of hypertension.[84]

11.5.1.2   COLD HOMOGENIZATION

Generally, cold homogenization is carried out when solid lipid is used; hence, high-pressure milling is required. Fine temperature and pressure control are required to ensure the unmolten state of lipids as homogenization process increases the temperature of the system and small exposure of biomolecules has been elevated to produce sustained activity.[85] Temperature-susceptible drug myricitrin was incorporated in this process and spherical nanoparticles were more effective as compared to rod shaped, playing a significant role of cold homogenization.[86]

The first preparation test is identical as in the hot homogenization, which includes the solubilization or dispersing of the drug in the melt of bulk lipid; however, the coming steps are different. The drug is first cooled inside the melt with the help of liquid nitrogen and high cooling rate governs homogeneous distribution. Cold homogenization generally results into larger particle size and higher polydispersity index (PDI) when compared to hot homogenization.[13]

11.5.2   EMULSIFICATION SONIFICATION

Synthesis of emulsion followed by sonication lead into nanoformulations of lipid with uniform distribution of narrow particle sizes. This is called emulsification–sonication method.[87] This method has various advantages including encapsulation of bioactives, like enzymes. The mixtures of phosphatidylcholine and cholesterol were dissolved in chloroform and film was formed using rotary evaporator. The buffer solution and lipid film were emulsified with sonication to obtain lipid vesicles.[88] Similarly, the radioactive substrate emulsion was prepared for encapsulating radiopharmaceutical using ultrasonic sonicator giving stable emulsions.[89]

11.5.3   MICROEMULSION-BASED LIPID NANOPARTICLES

The microemulsions are physicochemical unique and need exploration for a basic understanding of their formation, state of aggregation, internal interaction, and stability with reference to their probable uses. Their structural, dynamic, and transfer behavior was elaborated in a detailed way by Moulik and Pal.[90] The temperature parameter and value of pH control the product quality of microemulsion. The aggregation could be well prevented by altering the temperature gradients and also facilitates rapid lipid crystallization.[91]

11.5.4 *SOLVENT EMULSIFICATION/EVAPORATION*

Trotta et al. have shown the preparation method for nanoparticles based on the emulsification of butyl lactate or benzyl alcohol solution of a solid lipid in an aqueous solution. It was obtained by different emulsifiers, and thereafter followed by dilution of the emulsion with water and was used to prepare glyceryl monostearate nanodispersions with narrow size distribution.[58] One of the significances of this process over cold homogenization process is that stress on the lipid can be avoided and the use of organic solvent is a disadvantage.

11.5.5 *SOLVENT INJECTION*

Schubert and Goymann have shown the solvent injection as a new approach for manufacturing lipid nanoparticles.[36] It can be prepared by rapidly injecting a solution of solid lipids in water-miscible solvents or a water-miscible solvent mixture into water. The solvents used in this process are acetone, ethanol, isopropanol, and methanol, whereas ethyl acetate cannot produce the nanoparticles. Additionally, in a separate study, the lipid nanoparticles were produced using phospholipid as a lipid component and sucrose fatty acids as a surfactant with two different methods, namely, solvent injection and ultrasound emulsification. The study concluded with higher particle size using solvent injection when compared with ultrasound sonication in paclitaxel loaded lipid nanoparticles.[92] However, enhanced encapsulation efficiency up to 89% was reported with solvent injection method in paclitaxel-loaded lipid nanoparticles.[93]

11.5.6 *DOUBLE EMULSION*

W/o/w type of emulsions are prepared using suitable surfactant and potent lipid incorporated into the oil phase. The major significance of this method is prevention of drug partition to the external water during solvent evaporation during the formation of double emulsion.[55] Interestingly, it has been found to deliver the hydrophilic macromolecules orally like peptides and polypeptides.[37]

11.6   SOLID-STATE CHARACTERIZATION

11.6.1   PARTICLE SIZE AND POLYDISPERSITY INDEX

Laser diffraction (LD) and photon correlation spectroscopy are one of the most powerful analytical technique currently being used in the pharmaceutical research field to analyze the particle size and particle-size distribution. Photon correlation spectroscopy (PCS) is a dynamic light-scattering method, and it is one of the sensitive and accurate technique to measure very small particle sizes ranging from few nanometers to 3 μm.[94] PCS or quasi-elastic light scattering (QLS) is based on time-dependent fluctuations in the scattering intensity caused by small particles in suspension when a laser beam is applied to the sample (Fig. 11.1). Brownian motion of the dispersed particles causes the fluctuation in scattering intensity. Using the Stokes–Einstein equation and the data from these intensity fluctuations the hydrodynamic diameter of the particles can be calculated. However, this technique is unable to detect the particle sizes of larger microparticles. In these circumstances, the LD technique is mostly used.[95] LD is based on the phenomenon that laser beam is passed through a sample with dispersed particles the larger ones' scatter light at small angles while the smaller ones scatter light at large angles, which is referred as the Fraunhofer spectra (Fig. 11.2). LD methods sensitivity and coverage to detect a wide range of particle size ranging from nanometers to few millimeters make it preferred choice of the analytical technique. However, despite this advantage, it is recommended to use PCS and LD together to characterize SLNs. Use of both the methods can lead to more accurate information for colloidal systems with several particle populations can be determined. PCS and LD methods do not measure particle size directly. Instead, they use light diffraction and scattering to compute diameter, if the particles are spherical. For example, the LD method is not recommended for the particles, which are in nanometer size. However, this method gives a fast and accurate indication of the existence of microparticles. Fast results give some idea regarding the characteristics of microparticles and that may help formulators to decide or to make changes in formulation.

Particle-size distribution can be presented either as the PDI or by graphs. PDI equal to zero represents that the sample is monodisperse, and PDI close to one indicates that the sample is polydispersed.

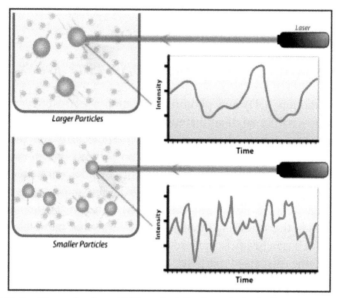

FIGURE 11.1   (See color insert.) The principle of quasi-elastic light scattering (QLS).

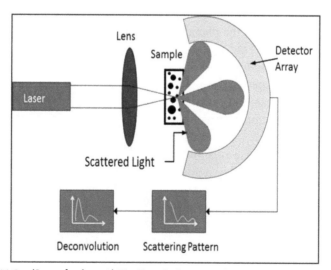

FIGURE 11.2   (See color insert.) The Fraunhofer spectra based on laser diffraction method.

11.6.2   ELECTRON MICROSCOPY

Difficulties may arise both in PCS and LD measurements for sample analysis due to crystallization; stress-induced morphological changes in SLNs of the sample have a wide range of particle-size distribution. Therefore, the use of an additional technique, such as scanning electron microscopy (SEM) and transmission electron microscopy (TEM), is recommended.[13] These techniques visualize the particles in nanometer ranges. The working of these instruments is based on the employment of electrons instead of light to provide particle images. TEM, as the name indicates, detects transmitted electrons and allows investigators to analyze the particle surface. While SEM detects the scattered electrons, which are scattered from the particle surface; this helps to visualize the particle and its surface. However, both techniques' results are highly dependent on the processing conditions, such as sample preparation, vacuum, and heating. Sometimes particles are coated or have a conductive surface that will cause a shiny surface which leads to imperfect particles images. Therefore, it is very important to control processing parameters by applying optimum vacuum and if samples are coated or highly conductive, then apply the gold coating. Application of gold coating may help to reduce the conductivity associated with the particle surface. SEM provides direct information on particle shape, size, and surface characteristics. This information helps to determine and characterize any morphological defects on the micro or nanoparticles that may alter the formulation properties in vivo. However, it is important to have a close observation of possible artifacts which may be the result of sample preparation. For example, the solvent evaporation method to prepare SLNs is widely used; however, the process of solvent removal may lead to some changes which will lead to changes in particle shape, size, or surface.

11.6.3   ATOMIC FORCE MICROSCOPY

Past decade atomic force microscopy (AFM) application have emerged in different areas such as of biotechnology, semiconductor devices, polymers, thin films, and mineral surface and the field of biological and pharmaceutical sciences.[96,97] This technique utilizes the force acting between surfaces and probing tip resulting in a spatial resolution of up to 0.01 nm for imaging. The biggest advantage of AFM over other technique is the simplicity in sample preparation. Unlike electron microscopy, AFM does not need the vacuum during the operation and the sample does not need to be conductive. Moreover, these advantages make AFM more suitable for the direct analysis of the originally hydrated, solvent-containing samples. However, there are some steps in sample preparation that needs to be taken care of such as particles immobilization. To assess particle shape by AFM, the particle must be immobilized and large particles show immobilization. However, particles which are very small in size move very fast. Therefore, some preparatory steps such as solvent removal are needed. Removal of the solvent increases the chances of change in the molecular structure which may not be ideal for sample analysis.

11.6.4   ZETA-POTENTIAL DETERMINATION

SLNs are the colloidal dispersion, and the stability of this dispersion can be demonstrated by measuring the zeta potential of a system.[7] The zeta potential gives information about the magnitude of the electrostatic repulsion or attraction between the particles in the aqueous suspension of SLNs. Zeta-potential value can serve as a very important aspect to establish the long-term stability of the formulation. Generally, high values of zeta-potential area are a good sign, and it can stabilize the colloidal system by repulsion. High zeta potential is more than 30 mV or less than –30 mV (Fig. 11.3). This electric repulsion between the particles results in less contact between particles. Less contact between the particles leads to no aggregation between the particles, and if there is no particle aggregation, then chances of seeing a change in particle morphology will be near to zero. However, this rule is not applicable when formulation has stabilizers. Formulation with stabilizers has zeta potential as low as 0 mV, still they show good long-term stability. Stabilizers will decrease the zeta potential due to the shift in the shear plane of the particle.[98]

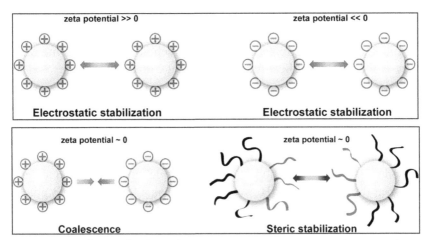

FIGURE 11.3 **(See color insert.)** Effect of zeta potential on particle attraction or repulsion.

11.6.5 CRYSTALLINITY AND CRYSTAL PACKING

Particle size, shape, and zeta-potential determination are necessary to characterize SLNs. However, these tests are not sufficient enough to characterize SLN's quality. Preparation of SLNs involves multiple steps and these steps may or may not lead to change in the lipid crystallinity and the modification of lipids. These crystalline parameters are closely related to the drug incorporation efficiency and release rate. Therefore, it is very important to characterize SLNs for their thermodynamic stability, crystal packing, and the degree of crystallinity.[34] Three major modifications occur during heating and cooling of lipids which are α-modification, β-modification, and β′-modification.[18] Besides these three major modifications, submodification exists. Cooling down SLN's formulation below the melting point of lipids is not always crystalline in nature. These particles are defined as a supercooled melt. Polymorphism and supercooled melts are two major problems associated with the production of SLNs. Therefore, it is very important to characterize the solid-state properties of SLNs. Solid-state characterization involves multiple techniques such as differential scanning calorimetry (DSC), powder X-ray diffraction (PXRD) studies, and vibrational spectroscopy.[99] Crystallinity and any crystalline changes are detected and determined by using DSC and PXRD.

DSC is a thermal analytical technique and widely used to determine crystallinity, polymorphism, phase-transition temperature and glass-forming ability of the material.[100,101] This technique measures the difference in the amount of heat required to change material state or to change material temperature compared to a reference. Negative or positive heat flow describes the nature of the process whether it is endothermic or exothermic. Different lipid modifications have different melting temperatures that could be easily detected by the use of DSC. The difference in melting points and melting enthalpies of different lipid modifications can provide detailed information about the sample structure and interaction between the components. However, if API has a higher melting point, it can dissolve in the melted lipid and tend to crystallize in lipid solid. This leads to false information about the lipid nature in SLNs.[102] Therefore, it is recommended to verify results from DSC using other orthogonal tests.

PXRD is a widely used analytical technique to determine materials' crystallinity, polymorphism, and subatomic distance.[103] In this technique, the sample is illuminated with X-rays of a fixed wavelength, and the intensity of the reflected radiation is recorded. It is necessary to have the sample in powder form; therefore, removal of water from colloidal dispersion is required. However, this technique may not be the best suit in analyzing samples in which crystals are formed as colloidal dispersion during storage.

X-ray scattering is another technique in which X-ray beams are passed through the sample obtained by a synchrotron.[104] Small angle X-ray and wide-angle X-ray are two techniques that fall under this category. Using these two techniques, colloidal suspensions can be characterized in their original state without any sample preparation. DSC and XRD is often supported by spectroscopic techniques such as infrared spectroscopy, Raman spectroscopy, electron spin resonance, and nuclear magnetic resonance.[105,106]

11.6.6   DRUG LOADING (DL%) AND ENTRAPMENT EFFICIENCY (EE%)

DL capacity and entrapment efficiency are the two most important parameters which are used to characterize how much drug is present in the formulated system.[107,108] The amount of the drug incorporated depends on the capacity of lipid to dissolve or disperse. Entrapment efficiency is the amount of the drug incorporated in the particles divided by its overall amount in the formulation. Entrapment efficiency is influenced by the characteristics of both the lipid and the API. DL capacity is another parameter that expresses the number of drug particles divided by the total carrier system. The most suitable technique used to determine drug concentration in the formulation or particles is liquid chromatography. Depending on the formulation, liquid chromatography is developed, such as high-performance liquid chromatography or ultraperformance liquid chromatography.

$$EE\% = \frac{\text{Amount of drug in the particles}}{\text{Amount of drug added in the formulation}} \times 100$$

$$DL\% = \frac{\text{Amount of drug in the particles}}{\text{Amount of drug added} + \text{Excipients added to the formulation}} \times 100$$

11.7   CONCLUSION

The LNs have unique applications in drug delivery system as a drug carrier. Its ability to improve the oral bioavailability and permeability makes it favorable drug delivery system for such drugs. The components used for fabrication of LNs are generally biocompatible and included in GRAS category. Various characterization techniques can be used for evaluation and optimization of LNs. Furthermore, LNs can be explored as a carrier for large macromolecules such as proteins and enzymes which leads to widen its scope.

KEYWORDS

- **solubility**
- **bioavailability**
- **lipidic nanoparticles**
- **drug delivery**

REFERENCES

1. Lindenberg, M.; Kopp, S.; Dressman, J. B. Classification of Orally Administered Drugs on the World Health Organization Model list of Essential Medicines According to the Biopharmaceutics Classification System. *Eur. J. Pharm. Biopharm.* **2004,** 58 (2), 265–278. https://www.sciencedirect.com/science/article/abs/pii/ S0939641104000438 (accessed Aug 18, 2018).

2. Attama, A. SLN, NLC, LDC: State of the Art in Drug and Active Delivery. Rec. Pat. Drug Deliv. Formul. **2011,** 5 (3), 178–187. http://www.eurekaselect.com/openurl/ content.php?genre=article&issn=1872-2113&volume=5&issue=3&spage=178 (accessed Aug 18, 2018).

3. Kesisoglou, F.; Panmai, S.; Wu, Y. Nanosizing—Oral Formulation Development and Biopharmaceutical Evaluation. Adv. Drug Deliv. Rev. **2007,** 59 (7), 631–644. https:// www.sciencedirect.com/science/article/pii/S0169409X0700083X (accessed Aug 18, 2018).

4. Mukherjee, S.; Ray, S.; Thakur, R. S. Solid Lipid Nanoparticles: A Modern Formulation Approach in Drug Delivery System. Indian J. Pharm. Sci. **2009,** 71 (4), 349–358. http://www.ncbi.nlm.nih.gov/pubmed/20502539 (accessed Aug 18, 2018).

5. Huang, L.-F.; Tong, W.-Q. Impact of Solid State Properties on Developability Assessment of Drug Candidates. Adv. Drug Deliv. Rev. **2004,** 56 (3), 321–334. https://www.sciencedirect.com/science/article/pii/S0169409X03002205 (accessed Aug 18, 2018).

6. Yu, L. X. Pharmaceutical Quality by Design: Product and Process Development, Understanding, and Control. Pharm. Res. **2008,** 25 (4), 781–791. http://link.springer. com/10.1007/s11095-007-9511-1 (accessed Aug 18, 2018).

7. Schwarz, C.; Mehnert, W.; Lucks, J. S.; Müller, R. H. Solid Lipid Nanoparticles (SLN) for Controlled Drug Delivery. I. Production, Characterization and Sterilization. J. Control. Release **1994,** 30 (1), 83–96. https://www.sciencedirect.com/science/article/ pii/0168365994900477 (accessed Aug 18, 2018).

8. Nastruzzi, C. Lipospheres in Drug Targets and Delivery: Approaches, Methods, and Applications; CRC Press: Boca Raton, FL, 2005; p 17. https://books.google.co.in/ books?hl=en&lr=&id=fdNLKrgt29IC&oi=fnd&pg=PA1&dq=acidan+lipid+sln&ots =YIxi-kop6S&sig=Ksgz8A5_mpDHvEEHLUK2REDbDz8#v=onepage&q=acidan lipid sln&f=false (accessed Aug 3, 2018).

9. Pardeike, J.; Hommoss, A.; Müller, R. H. Lipid Nanoparticles (SLN, NLC) in Cosmetic and Pharmaceutical Dermal Products. Int. J. Pharm. **2009,** 366 (1–2), 170–184.

10. Müller, R. Solid Lipid Nanoparticles (SLN) for Controlled Drug Delivery: A Review of the State of the Art. Eur. J. Pharm. Biopharm. **2000,** 50 (1), 161–177.

11. Mehnert, W.; Mäder, K. Solid Lipid Nanoparticles: Production, Characterization and Applications. Adv. Drug Deliv. Rev. **2001,** 47 (2–3), 165–196. https://www. sciencedirect.com/science/article/pii/S0169409X01001053?via%3Dihub (accessed Aug 18, 2018).

12. Li, H.; Zhao, X.; Ma, Y.; Zhai, G.; Li, L.; Lou, H. Enhancement of Gastrointestinal Absorption of Quercetin by Solid Lipid Nanoparticles. J. Control. Release

2009, 133 (3), 238–244. https://www.sciencedirect.com/science/article/pii/ S0168365908006299 (accessed Aug 18, 2018).

13. Mehnert, W.; Mäder, K. Solid Lipid Nanoparticles: Production, Characterization and Applications. Adv. Drug Deliv. Rev. **2012,** 64 (Suppl.), 83–101. http://dx.doi. org/10.1016/j.addr.2012.09.021, accessed Aug 3, 2018.

14. Cavalli, R.; Caputo, O.; Eugenia, M.; Trotta, M.; Scarnecchia, C.; Gasco, M. R. Sterilization and Freeze-Drying of Drug-Free and Drug-Loaded Solid Lipid Nanoparticles. Int. J. Pharm. **1997,** 148 (1), 47–54.

15. Torchilin, V.; Vladimir, P.; Amiji, M. M. Handbook of Materials for Nanomedicine; Pan Stanford Publishing: India; 2011; p 855. https://books.google.co.in/books?hl= en&lr=&id=o8HgyfvvkzIC&oi=fnd&pg=PA383&dq=acidan+lipid+sln&ots=OZi PVe5AnR&sig=KXbBjLBoc_JpWY_slhbpbZsi9Kw#v=onepage&q=acidan   lipid sln&f=false (accessed Aug 3, 2018).

16. Jenning, V.; Mäder, K.; Gohla, S. H. Solid Lipid Nanoparticles (SLNTM) Based on Binary Mixtures of Liquid and Solid Lipids: A 1H-NMR Study. Int. J. Pharm. **2000,** 205 (1–2), 15–21. https://www.sciencedirect.com/science/article/pii/ S0378517300004622 (accessed Aug 3, 2018).

17. Puglia, C.; Blasi, P.; Rizza, L.; Schoubben, A.; Bonina, F.; Rossi, C.; et al. Lipid Nanoparticles for Prolonged Topical Delivery: An In Vitro and In Vivo Investigation. Int. J. Pharm. **2008,** 357 (1–2), 295–304. https://www.sciencedirect.com/science/ article/pii/S0378517308000951 (accessed Aug 3, 2018).

18. Freitas, C.; Müller, R. H. Correlation between Long-Term Stability of Solid Lipid Nanoparticles (SLNTM) and Crystallinity of the Lipid Phase. Eur. J. Pharm. Biopharm. **1999,** 47 (2), 125–132. https://www.sciencedirect.com/science/article/pii/ S0939641198000745 (accessed Aug 3, 2018).

19. Cavalli, R.; Morel, S.; Gasco, M. Preparation and Evaluation In Vitro of Colloidal Liphospheres Containing Pilocarpine as Ion Pair. Int. J. Pharm. **1995,** 117 (2), 243–246. https://www.sciencedirect.com/science/article/pii/0378517394003397   (accessed Aug 3, 2018).

20. Morel, S.; Rosa Gasco, M.; Cavalli, R. Incorporation in Liphospheres of [d-Trp-6] LHRH. Int. J. Pharm. **1994,** 105 (2), R1–R3. https://www.sciencedirect.com/science/ article/pii/0378517394904669 (accessed Aug 3, 2018).

21. Ugazio, E.; Cavalli, R.; Gasco, M. R. Incorporation of Cyclosporin A in Solid Lipid Nanoparticles (SLN). Int. J. Pharm. **2002,** 241 (2), 341–344. https://www. sciencedirect.com/science/article/pii/S0378517302002685 (accessed Aug 3, 2018).

22. Wong, H. L.; Bendayan, R.; Rauth, A. M.; Li, Y.; Wu, X. Y. Chemotherapy with Anticancer Drugs Encapsulated in Solid Lipid Nanoparticles. Adv. Drug Deliv. Rev. **2007,** 59 (6), 491–504. https://www.sciencedirect.com/science/article/pii/ S0169409X07000440 (accessed Aug 3, 2018).

23. Maier, M. A.; Jayaraman, M.; Matsuda, S.; Liu, J., Barros, S.; Querbes, W.; et al. Biodegradable Lipids Enabling Rapidly Eliminated Lipid Nanoparticles for Systemic Delivery of RNAi Therapeutics. Mol. Ther. **2013,** 21 (8), 1570–1578. https://www. sciencedirect.com/science/article/pii/S1525001616319864 (accessed Aug 3, 2018).

24. Teeranachaideekul, V.; Souto, E. B.; Junyaprasert, V. B.; Müller, R. H. Cetyl Palmitate-Based NLC for Topical Delivery of Coenzyme Q10—Development, Physicochemical Characterization and In Vitro Release Studies. Eur. J. Pharm.

Biopharm. **2007,** 67 (1), 141–148. https://www.sciencedirect.com/science/article/pii/ S0939641107000161 (accessed Aug 3, 2018).

25. Volkhard, J.; Sven, H.; Gohla, S. H. G. Encapsulation of Retinoids in Solid Lipid Nanoparticles (SLN). J. Microencapsul. **2001,** 18 (2), 149–158. http://www.tandfonline.com/doi/full/10.1080/02652040010000361 (accessed Aug 3, 2018).

26. Müller, R. H.; Radtke, M.; Wissing, S. A. Solid Lipid Nanoparticles (SLN) and Nanostructured Lipid Carriers (NLC) in Cosmetic and Dermatological Preparations. Adv. Drug Deliv. Rev. **2002,** 54, S131–S155. https://www.sciencedirect.com/science/article/pii/S0169409X02001187 (accessed Aug 3, 2018).

27. Gasco, M. R. Method for Producing Solid Lipid Microspheres Having a Narrow Size Distribution. US5250236A, 1991. Available from: https://patents.google.com/patent/US5250236A/en (accessed Aug 3, 2018).

28. Li, Y.; Taulier, N.; Rauth, A. M.; Wu, X. Y. Screening of Lipid Carriers and Characterization of Drug–Polymer–Lipid Interactions for the Rational Design of Polymer–Lipid Hybrid Nanoparticles (PLN). Pharm. Res. **2006,** 23 (8), 1877–1887. http://link.springer.com/10.1007/s11095-006-9033-2 (accessed Aug 3, 2018).

29. Zhang, L.; Pornpattananangkul, D.; Hu, C.-M.; Huang, C.-M. Development of Nanoparticles for Antimicrobial Drug Delivery. Curr. Med. Chem. **2010,** 17 (6), 585–594.

30. Ryde, N. P.; Ruddy, S. B. Solid Dose Nanoparticulate Compositions Comprising a Synergistic Combination of a Polymeric Surface Stabilizer and Dioctyl Sodium Sulfosuccinate. US6375986B1, 2000.

31. Tiyaboonchai, W.; Tungpradit, W.; Plianbangchang, P. Formulation and Characterization of Curcuminoids Loaded Solid Lipid Nanoparticles. Int. J. Pharm. **2007,** 337 (1–2), 299–306.

32. Na, G. C.; Rajagopalan, N. Method of Preparing Nanoparticle Compositions Containing Charged Phospholipids to Reduce Aggregation. United States Patent 5470583, 1994.

33. Sznitowska, M.; Gajewska, M.; Janicki, S.; Radwanska, A.; Lukowski, G. Bioavailability of Diazepam from Aqueous-Organic Solution, Submicron Emulsion and Solid Lipid Nanoparticles after Rectal Administration in Rabbits. Eur. J. Pharm. Biopharm. **2001,** 52 (2), 159–163.

34. Bunjes, H.; Koch, M. H. J. Saturated Phospholipids Promote Crystallization But Slow Down Polymorphic Transitions in Triglyceride Nanoparticles. J. Control. Release **2005,** 107 (2), 229–243.

35. Dudhipala, N.; Veerabrahma, K. Candesartan Cilexetil Loaded Solid Lipid Nanoparticles for Oral Delivery: Characterization, Pharmacokinetic and Pharmacodynamic Evaluation. Drug Deliv. **2016,** 23 (2), 395–404.

36. Schubert, M. A.; Müller-Goymann, C. C. Solvent Injection as a New Approach for Manufacturing Lipid Nanoparticles—Evaluation of the Method and Process Parameters. Eur. J. Pharm. Biopharm. **2003,** 55 (1), 125–131.

37. Silva, A. C.; González-Mira, E.; García, M. L.; Egea, M. A.; Fonseca, J.; Silva, R.; et al. Preparation, Characterization and Biocompatibility Studies on Risperidone-Loaded Solid Lipid Nanoparticles (SLN), High Pressure Homogenization Versus Ultrasound. Colloids Surf. B: Biointerfaces **2011,** 86 (1), 158–165.

38. Wissing, S. A.; Müller, R. H. The Influence of Solid Lipid Nanoparticles on Skin Hydration and Viscoelasticity—In Vivo Study. Eur. J. Pharm. Biopharm. **2003,** 56 (1), 67–72.

39. Abbasalipourkabir, R.; Salehzadeh, A.; Abdullah, R. Solid Lipid Nanaoparticles as New Drug Delivery System. Int. J. Biotechnol. Mol. Biol. Res. **2011,** 2 (13), 252–261.

40. Zhang, R. X.; Ahmed, T.; Li, L. Y.; Li, J.; Abbasi, A. Z.; Wu, X. Y. Design of Nanocarriers for Nanoscale Drug Delivery to Enhance Cancer Treatment Using Hybrid Polymer and Lipid Building Blocks. Nanoscale **2017,** 9 (4), 1334–1355.

41. Xie, S.; Zhu, L.; Dong, Z.; Wang, Y.; Wang, X.; Zhou, W. Preparation and Evaluation of Ofloxacin-Loaded Palmitic Acid Solid Lipid Nanoparticles. Int. J. Nanomed. **2011,** 6, 547–555.

42. Xie, S.; Zhu, L.; Dong, Z.; Wang, X.; Wang, Y.; Li, X.; et al. Preparation, Characterization and Pharmacokinetics of Enrofloxacin-Loaded Solid Lipid Nanoparticles: Influences of Fatty Acids. Colloids Surf. B: Biointerfaces **2011,** 83 (2), 382–387.

43. Liu, J.; Gong, T.; Wang, C.; Zhong, Z.; Zhang, Z. Solid Lipid Nanoparticles Loaded with Insulin by Sodium Cholate-Phosphatidylcholine-Based Mixed Micelles: Preparation and Characterization. Int. J. Pharm. **2007,** 340 (1–2), 153–162.

44. Rasch, M. R.; Rossinyol, E.; Hueso, J. L.; Goodfellow, B. W.; Arbiol, J.; Korgel, B. A. Hydrophobic Gold Nanoparticle Self-Assembly with Phosphatidylcholine Lipid: Membrane-Loaded and Janus Vesicles. Nano Lett. **2010,** 10 (9), 3733–3739.

45. Venkateswarlu, V.; Manjunath, K. Preparation, Characterization and In Vitro Release Kinetics of Clozapine Solid Lipid Nanoparticles. J. Control. Release **2004,** 95 (3), 627–638.

46. Sahay, G.; Querbes, W.; Alabi, C.; Eltoukhy, A.; Sarkar, S.; Zurenko, C.; et al. Efficiency of siRNA Delivery by Lipid Nanoparticles Is Limited by Endocytic Recycling. Nat. Biotechnol. **2013,** 31 (7), 653–658.

47. Hu, F.-Q.; Jiang, S.-P.; Du, Y.-Z.; Yuan, H.; Ye, Y.-Q.; Zeng, S. Preparation and Characterization of Stearic Acid Nanostructured Lipid Carriers by Solvent Diffusion Method in an Aqueous System. Colloids Surf. B: Biointerfaces **2005,** 45 (3–4), 167–173.

48. Jójárt, B.; Poša, M.; Fiser, B.; Szőri, M.; Farkaš, Z.; Viskolcz, B. Mixed Micelles of Sodium Cholate and Sodium Dodecylsulphate 1:1 Binary Mixture at Different Temperatures—Experimental and Theoretical Investigations. PLoS One **2014,** 9 (7), e102114.

49. Mudshinge, S. R.; Deore, A. B.; Patil, S.; Bhalgat, C. M. Nanoparticles: Emerging Carriers for Drug Delivery. Saudi Pharm. J. **2011,** 19 (3), 129–141.

50. Westesen, K.; Bunjes, H.; Koch, M. H. Physicochemical Characterization of Lipid Nanoparticles and Evaluation of Their Drug Loading Capacity and Sustained Release Potential. J. Control. Release **1997,** 48 (2–3), 223–236.

51. Esposito, E.; Fantin, M.; Marti, M.; Drechsler, M.; Paccamiccio, L.; Mariani, P.; et al. Solid Lipid Nanoparticles as Delivery Systems for Bromocriptine. Pharm. Res. **2008,** 25 (7), 1521–1530.

52. Lim, S.-J.; Kim, C.-K. Formulation Parameters Determining the Physicochemical Characteristics of Solid Lipid Nanoparticles Loaded with All-trans Retinoic Acid. Int. J. Pharm. **2002,** 243 (1–2), 135–146.

53. Lee, M.-K.; Lim, S.-J., Kim, C.-K. Preparation, Characterization and In Vitro Cytotoxicity of Paclitaxel-Loaded Sterically Stabilized Solid Lipid Nanoparticles. Biomaterials 2007, 28 (12), 2137–2146.

54. Martins, S.; Silva, A. C.; Ferreira, D. C.; Souto, E. B. Improving Oral Absorption of Samon Calcitonin by Trimyristin Lipid Nanoparticles. J. Biomed. Nanotechnol. 2009, 5 (1), 76–83.

55. Wissing, S.; Kayser, O.; Müller, R. Solid Lipid Nanoparticles for Parenteral Drug Delivery. Adv. Drug Deliv. Rev. 2004, 56 (9), 1257–1272.

56. Trombino, S.; Cassano, R.; Muzzalupo, R.; Pingitore, A.; Cione, E.; Picci, N. Stearyl Ferulate-Based Solid Lipid Nanoparticles for the Encapsulation and Stabilization of β-Carotene and α-Tocopherol. Colloids Surf. B: Biointerfaces 2009, 72 (2), 181–187.

57. Patlolla, R. R.; Chougule, M.; Patel, A. R.; Jackson, T.; Tata, P. N. V.; Singh, M. Formulation, Characterization and Pulmonary Deposition of Nebulized Celecoxib Encapsulated Nanostructured Lipid Carriers. J. Control. Release 2010, 144 (2), 233–241.

58. Trotta, M.; Debernardi, F.; Caputo, O. Preparation of Solid Lipid Nanoparticles by a Solvent Emulsification–Diffusion Technique. Int. J. Pharm. 2003, 257 (1–2), 153–160.

59. Gallarate, M.; Trotta, M.; Battaglia, L.; Chirio, D. Preparation of Solid Lipid Nanoparticles from W/O/W Emulsions: Preliminary Studies on Insulin Encapsulation. J. Microencapsul. 2009, 26 (5), 394–402.

60. Schwarz, C. Solid Lipid Nanoparticles (SLN) for Controlled Drug Delivery. II. Drug Incorporation and Physicochemical Characterization. J. Microencapsul. 1999, 16 (2), 205–213.

61. Üner, M.; Wissing, S. A.; Yener, G.; Müller, R. H. Skin Moisturizing Effect and Skin Penetration of Ascorbyl Palmitate Entrapped in Solid Lipid Nanoparticles (SLN) and Nanostructured Lipid Carriers (NLC) incorporated into hydrogel. Pharmazie 2005, 60 (10), 751–755.

62. Mandawgade, S. D.; Patravale, V. B. Development of SLNs from Natural Lipids: Application to Topical Delivery of Tretinoin. Int. J. Pharm. 2008, 363 (1–2), 132–138.

63. Joshi, M.; Patravale, V. Formulation and Evaluation of Nanostructured Lipid Carrier (NLC)-based Gel of Valdecoxib. Drug Dev. Ind. Pharm. 2006, 32 (8), 911–918.

64. Müller, R.; Maaben, S.; Weyhers, H.; Mehnert, W. Phagocytic Uptake and Cytotoxicity of Solid Lipid Nanoparticles (SLN) Sterically Stabilized with Poloxamine 908 and Poloxamer 407. J. Drug Target. 1996, 4 (3), 161–170.

65. Sanjula, B.; Shah, F. M.; Javed, A.; Alka, A. Effect of Poloxamer 188 on Lymphatic Uptake of Carvedilol-Loaded Solid Lipid Nanoparticles for Bioavailability Enhancement. J. Drug Target. 2009, 17 (3), 249–256.

66. Martins, S.; Tho, I.; Souto, E.; Ferreira, D.; Brandl, M. Multivariate Design for the Evaluation of Lipid and Surfactant Composition Effect for Optimisation of Lipid Nanoparticles. Eur. J. Pharm. Sci. 2012, 45 (5), 613–623.

67. Anurak, L.; Chansiri, G.; Peankit, D.; Somlak, K. Griseofulvin Solid Lipid Nanoparticles Based on Microemulsion Technique. Adv. Mater. Res. 2011, 197–198, 47–50.

68. Hu, F.; Yuan, H.; Zhang, H.; Fang, M. Preparation of Solid Lipid Nanoparticles with Clobetasol Propionate by a Novel Solvent Diffusion Method in Aqueous System and Physicochemical Characterization. Int. J. Pharm. **2002,** 239 (1–2), 121–128.
69. Hu, F.; Hong, Y.; Yuan, H. Preparation and Characterization of Solid Lipid Nanoparticles Containing Peptide. Int. J. Pharm. **2004,** 273 (1–2), 29–35.
70. Müller, R. H.; Runge, S.; Ravelli, V.; Mehnert, W.; Thünemann, A. F.; Souto, E. B. Oral Bioavailability of Cyclosporine: Solid Lipid Nanoparticles (SLN®) Versus Drug Nanocrystals. Int. J. Pharm. **2006,** 317 (1), 82–89.
71. Yang, S. C.; Lu, L. F.; Cai, Y.; Zhu, J. B.; Liang, B. W.; Yang, C. Z. Body Distribution in Mice of Intravenously Injected Camptothecin Solid Lipid Nanoparticles and Targeting Effect on Brain. J. Control. Release **1999,** 59 (3), 299–307.
72. Nayak, A. P.; Tiyaboonchai, W.; Patankar, S.; Madhusudhan, B.; Souto, E. B. Curcuminoids-Loaded Lipid Nanoparticles: Novel Approach towards Malaria Treatment. Colloids Surf. B: Biointerfaces **2010,** 81 (1), 263–273.
73. Pawar, K. R.; Babu, R. J. Lipid Materials for Topical and Transdermal Delivery of Nanoemulsions. Crit. Rev. Ther. Drug Carrier Syst. **2014,** 31 (5), 429–458.
74. Khatak, S.; Dureja, H. Recent Techniques and Patents on Solid Lipid Nanoparticles as Novel Carrier for Drug Delivery. Recent Pat Nanotechnol. **2015,** 9 (3), 150–177.
75. Shah, K. A.; Date, A. A.; Joshi, M. D.; Patravale, V. B. Solid Lipid Nanoparticles (SLN) of Tretinoin: Potential in Topical Delivery. Int. J. Pharm. **2007,** 345 (1–2), 163–171.
76. Bhalekar, M. R.; Pokharkar, V.; Madgulkar, A.; Patil, N.; Patil, N. Preparation and Evaluation of Miconazole Nitrate-Loaded Solid Lipid Nanoparticles for Topical Delivery. AAPS PharmSciTech **2009,** 10 (1), 289–296.
77. Göppert, T. M.; Müller, R. H. Protein Adsorption Patterns on Poloxamer- and Poloxamine-Stabilized Solid Lipid Nanoparticles (SLN). Eur. J. Pharm. Biopharm. **2005,** 60 (3), 361–372.
78. Kheradmandnia, S.; Vasheghani-Farahani, E.; Nosrati, M.; Atyabi, F. Preparation and Characterization of Ketoprofen-Loaded Solid Lipid Nanoparticles Made from Beeswax and Carnauba Wax. Nanomed. Nanotechnol., Biol. Med. **2010,** 6 (6), 753–759.
79. Choi, S.-Y.; Oh, S.-G.; Bae, S.-Y.; Moon, S.-K. Effect of Short-Chain Alcohols as Co-surfactants on Pseudo-ternary Phase Diagrams Containing Lecithin. Korean J. Chem. Eng. **1999,** 16 (3), 377–381. http://link.springer.com/10.1007/BF02707128 (accessed Aug 1, 2018).
80. Salminen, H.; Helgason, T.; Aulbach, S.; Kristinsson, B.; Kristbergsson, K.; Weiss, J. Influence of Co-surfactants on Crystallization and Stability of Solid Lipid Nanoparticles. J. Colloid Interface Sci. **2014,** 426, 256–263. http://dx.doi.org/10.1016/j.jcis.2014.04.009, accessed Aug 3, 2018.
81. Floury, J.; Desrumaux, A.; Lardières, J. Effect of High-Pressure Homogenization on Droplet Size Distributions and Rheological Properties of Model Oil-in-Water Emulsions. Innov. Food Sci. Emerg. Technol. **2000,** 1 (2), 127–134.
82. Sevenich, R.; Mathys, A. Continuous Versus Discontinuous Ultra-High-Pressure Systems for Food Sterilization with Focus on Ultra-High-Pressure Homogenization and High-Pressure Thermal Sterilization: A Review. Compr. Rev. Food Sci. Food Saf. **2018,** 17 (3), 646–662.

83. Lander, R.; Manger, W.; Scouloudis, M.; Ku, A.; Davis, C.; Lee, A. Gaulin Homogenization: A Mechanistic Study. Biotechnol. Prog. **2000**, 16 (1), 80–85. http://www.ncbi.nlm.nih.gov/pubmed/10662494 (accessed Aug 6, 2018).

84. Pandya, N. T.; Jani, P.; Vanza, J.; Tandel, H. Solid Lipid Nanoparticles as an Efficient Drug Delivery System of Olmesartan Medoxomil for the Treatment of Hypertension. Colloids Surf. B: Biointerfaces **2018**, 165, 37–44. https://www.sciencedirect.com/science/article/pii/S0927776518300808 (accessed Aug 6, 2018).

85. Sharma, G.; Chopra, K.; Puri, S.; Bishnoi, M.; Rishi, P.; Kaur, I. P. Topical Delivery of TRPsiRNA-Loaded Solid Lipid Nanoparticles Confer Reduced Pain Sensation via TRPV1 Silencing, in Rats. J. Drug Target. **2018**, 26 (2), 135–149. https://www.tandfonline.com/doi/full/10.1080/1061186X.2017.1350857 (accessed Aug 6, 2018).

86. Ahangarpour, A.; Oroojan, A. A.; Khorsandi, L.; Kouchak, M.; Badavi, M. Solid Lipid Nanoparticles of Myricitrin Have Antioxidant and Antidiabetic Effects on Streptozotocin-Nicotinamide-Induced Diabetic Model and Myotube Cell of Male Mouse. Oxid. Med. Cell Longev. **2018**, 2018, 1–18. https://www.hindawi.com/journals/omcl/2018/7496936/ (accessed Aug 6, 2018).

87. Luo, Y.; Chen, D.; Ren, L.; Zhao, X.; Qin, J. Solid Lipid Nanoparticles for Enhancing Vinpocetine's Oral Bioavailability. J. Control. Release **2006**, 114 (1), 53–59. https://www.sciencedirect.com/science/article/pii/S0168365906002203 (accessed Aug 6, 2018).

88. Papahadjopoulos, D. P.; Szoka, Jr., F. C. Method of Encapsulating Biologically Active Materials in Lipid Vesicles. US4235871A, 1978.

89. Nilsson-Ehle, P.; Schotz, M. C. A Stable, Radioactive Substrate Emulsion for Assay of Lipoprotein Lipase. J. Lipid Res. **1976**, 17 (5), 536–541.

90. Moulik, S. P.; Paul, B. K. Structure, Dynamics and Transport Properties of Microemulsions. Adv. Colloid Interface Sci. **1998**, 78 (2), 99–195.

91. Zimmermann, E.; Müller, R.; Mäder, K. Influence of Different Parameters on Reconstitution of Lyophilized SLN. Int. J. Pharm. **2000**, 196 (2), 211–213.

92. Arıca Yegin, B.; Benoît, J.-P.; Lamprecht, A. Paclitaxel-Loaded Lipid Nanoparticles Prepared by Solvent Injection or Ultrasound Emulsification. Drug Dev. Ind. Pharm. **2006**, 32 (9), 1089–1094.

93. Pandita, D.; Ahuja, A.; Velpandian, T.; Lather, V.; Dutta, T.; Khar, R. K. Characterization and In Vitro Assessment of Paclitaxel Loaded Lipid Nanoparticles Formulated Using Modified Solvent Injection Technique. Pharmazie **2009**, 64 (5), 301–310.

94. Jores, K.; Mehnert, W.; Drechsler, M.; Bunjes, H.; Johann, C.; Mäder, K. Investigations on the Structure of Solid Lipid Nanoparticles (SLN) and Oil-Loaded Solid Lipid Nanoparticles by Photon Correlation Spectroscopy, Field-Flow Fractionation and Transmission Electron Microscopy. J. Control. Release **2004**, 95 (2), 217–227. https://www.sciencedirect.com/science/article/pii/S016836590300556X (accessed Aug 18, 2018).

95. Shekunov, B. Y.; Chattopadhyay, P.; Tong, H. H. Y.; Chow, A. H. L. Particle Size Analysis in Pharmaceutics: Principles, Methods and Applications. Pharm. Res. **2007**, 24 (2), 203–227. http://link.springer.com/10.1007/s11095-006-9146-7 (accessed Aug 18, 2018).

96. Ruozi, B.; Tosi, G.; Forni, F.; Fresta, M.; Vandelli, M. A. Atomic Force Microscopy and Photon Correlation Spectroscopy: Two Techniques for Rapid Characterization of Liposomes. Eur. J. Pharm. Sci. **2005**, 25 (1), 81–89. https://www.sciencedirect. com/science/article/pii/S0928098705000606?via%3Dihub (accessed Aug 18, 2018).

97. Braga, P. C.; Ricci, D. Atomic Force Microscopy in Biomedical Research: Methods and Protocols; Humana: New York, 2011; 508 p.

98. Heurtault, B.; Saulnier, P.; Pech, B.; Proust, J.-E.; Benoit, J.-P. Physico-chemical Stability of Colloidal Lipid Particles. Biomaterials **2003**, 24 (23), 4283–4300. https:// www.sciencedirect.com/science/article/pii/S0142961203003314?via%3Dihub (accessed Aug 18, 2018).

99. Vippagunta, S. R.; Brittain, H. G.; Grant, D. J. W. Crystalline Solids. Adv. Drug Deliv. Rev. **2001**, 48 (1), 3–26. https://www.sciencedirect.com/science/article/pii/ S0169409X01000977?via%3Dihub (accessed Aug 18, 2018).

100. Clas, S. D.; Dalton, C. R.; Hancock, B. C. Differential Scanning Calorimetry: Applications in Drug Development. Pharm. Sci. Technol. Today **1999**, 2 (8), 311–320. http://www.ncbi.nlm.nih.gov/pubmed/10441275 (accessed Aug 18, 2018).

101. Demetzos, C. Differential Scanning Calorimetry (DSC): A Tool to Study the Thermal Behavior of Lipid Bilayers and Liposomal Stability. J Liposome Res. **2008**, 18 (3), 159–173. http://www.tandfonline.com/doi/full/10.1080/08982100802310261 (accessed Aug 18, 2018).

102. Hancock, B. C.; Zografi, G. Characteristics and Significance of the Amorphous State in Pharmaceutical Systems. J. Pharm. Sci. **1997**, 86 (1), 1–12. http://linkinghub. elsevier.com/retrieve/pii/S002235491550227X (accessed Aug 18, 2018).

103. Svilenov, H.; Tzachev, C. Solid Lipid Nanoparticles—A Promising Drug Delivery. Nanomedicine **2009**, 2, 187–237.

104. Dong, Y.-D.; Boyd, B. J. Applications of X-ray Scattering in Pharmaceutical Science. Int. J. Pharm. **2011**, 417 (1–2), 101–111. https://www.sciencedirect.com/science/ article/pii/S0378517311000536 (accessed Aug 18, 2018).

105. Westesen, K.; Siekmann, B.; Koch, M. H. J. Investigations on the Physical State of Lipid Nanoparticles by Synchrotron Radiation X-ray Diffraction. Int. J. Pharm. **1993**, 93 (1–3), 189–199. http://linkinghub.elsevier.com/retrieve/pii/037851739390177H (accessed Aug 18, 2018).

106. Li, Y.; Chow, P. S.; Tan, R. B. H. Quantification of Polymorphic Impurity in an Enantiotropic Polymorph System Using Differential Scanning Calorimetry, X-ray Powder Diffraction and Raman Spectroscopy. Int. J. Pharm. **2011**, 415 (1–2), 110–118. https://www.sciencedirect.com/science/article/pii/ S0378517311005060?via%3Dihub (accessed Aug 18, 2018)

107. Bugay, D. E. Characterization of the Solid-State: Spectroscopic Techniques. Adv. Drug Deliv. Rev. **2001**, 48 (1), 43–65. https://www.sciencedirect.com/science/article/ pii/S0169409X01001016?via%3Dihub (accessed Aug 16, 2018)

108. Bala, I.; Hariharan, S.; Kumar, M. N. V. R. PLGA Nanoparticles in Drug Delivery: The State of the Art. Crit. Rev. Ther. Drug Carrier Syst. **2004**, 21 (5), 387–422. http:// www.ncbi.nlm.nih.gov/pubmed/15719481 (accessed Aug 18, 2018)

CHAPTER 12

MANHATTAN PROJECT, *ATOMS FOR PEACE*, NUCLEAR WEAPONS, AND ACCIDENTS

FRANCISCO TORRENS[1,*] and GLORIA CASTELLANO[2]

[1]*Institut Universitari de Ciència Molecular, Universitat de València, Edifici d'Instituts de Paterna, PO Box 22085, E-46071 València, Spain*

[2]*Departamento de Ciencias Experimentales y Matemáticas, Facultad de Veterinaria y Ciencias Experimentales, Universidad Catolica de Valencia San Vicente Mártir, Guillem de Castro-94, E-46001 València, Spain*

Corresponding author. E-mail: torrens@uv.es

ABSTRACT

Nuclear science modernity contrasts with the reactionary ideology of Franco system. There is no necessary historical relationship between science and democracy. Manhattan Project was conceived after a letter, in which US President Roosevelt was urged to begin nuclear program, which was outlined by Szilard and signed by Einstein. Influenced by Oppenheimer's pacifist ideas, US President Eisenhower made the speech *Atoms for Peace* in the seat of the United Nations. After Eisenhower delivered *Atoms for Peace* speech to the United Nations General Assembly and Atomic Energy Commission Programme, Spanish traveling show presented peaceful applications of nuclear energy to general public, with materials coming from Atomic Energy Commission Programme. American nuclear cover-up in Spain occurred after Palomares disaster. Making of ignorance versus conscience in Palomares accident happened via official versus local speeches. Peace Boat is a Japan-based, international nongovernmental organization that works to promote peace and sustainability via organization of peace voyages onboard a large passenger ship. There is a parallelism among Chernobyl, Fukushima, and Cofrents nuclear power station. An energy model is possible without risks, clean, economic, sustainable, renewable, distributed, democratic, and generating stable and quality employment.

12.1 INTRODUCTION

Nuclear science modernity contrasts with the reactionary ideology of Franco system (1939–1975). There is no necessary historical relationship between science and democracy.[1]

Manhattan Project (MP, 1942–1947) was conceived after a letter, in which US President Roosevelt was urged to begin a nuclear program, which was outlined by Szilard and signed by Einstein (1939). Influenced by Oppenheimer's pacifist ideas, US President Eisenhower (1953), old general that fought in World War II (WW2) in Europe, made the speech *Atoms for Peace* (AFP) in the seat of United Nations (UN). After Eisenhower delivered AFP speech to the UN General Assembly and Atomic Energy Commission (AEC) Programme, Spanish traveling show (1958) presented peaceful applications of nuclear energy to general public, with materials coming from AEC Programme. American nuclear cover-up in Spain occurred after Palomares (Almería) disaster (1966). Making of ignorance versus conscience in Palomares accident happened via official versus local speeches. The objective is to reconstruct the processes of making of ignorance via the campaign of making of ignorance. Peace Boat is a Japan-based, international nongovernmental organization (NGO) that works since 1983 to promote peace and sustainability via organization of peace voyages onboard a large passenger ship. Activities must be based on the philosophy that any problem faced by any community is a global challenge that must be tackled via cooperation between people, organizations, and governments of the world. There is a parallelism among Chernobyl, Fukushima, and Cofrents (València, Spain) nuclear power station (1984–). An energy model is possible without risks, clean, economic, sustainable, renewable, distributed, democratic, and generating stable and quality employment.

In earlier publications, nuclear fusion and American nuclear cover-up in Spain after Palomares (Almería) disaster (1966) were informed.[2] The aim of this work is to initiate a debate by suggesting a number of questions (Q), which can arise when addressing subjects of AFP, in different fields, and providing, when possible, answers (A), hypotheses (H), and paradoxes (P). Specific fields of the debate are as follows: American nuclear cover-up in Spain after Palomares disaster; making of ignorance versus conscience in Palomares accident via official versus local speeches; radiography of nuclear weapons; and Chernobyl, Fukushima, and the nuclear power station of Cofrents (València, Spain, 1984).

12.2 MANHATTAN PROJECT (1942–1947)

The MP (1942–1947) was conceived after a letter, in which US President Roosevelt was urged to begin a nuclear program. The letter was outlined by Szilard and signed by Einstein (1939, cf. Fig. 12.1).[3] Both scientists of Jewish origin passed by a moment of panic when they got a slight idea of nuclear energy power, and National-Socialist (*Nationalsozialistische*, Nazi) Germany research [Uranium Project (1939–1945) headed by Heisenberg]. Both believed that if nuclear energy reached German Reich, it could radically change the course of WW2. In an exonerating way, both did not already know if Germans were concentrating their research efforts on nuclear energy or other fields. Really, exertions were not focused on Uranium Project because it did not promise results before WW2 end.[4] With hindsight, after Cold War began and Armageddon of humanity was seen as possible and probable, Einstein said that *I could burn my fingers that I wrote that letter*, which was written in 1939, and MP began in 1942, counting on many greatest and most brilliant minds of the time.[3] He did not work in MP as his pacifist tendencies were considered a risk of security.

Scientists that worked in MP were famous and brilliant (Fermi, Feynman, Bethe, Oppenheimer, cf. Fig. 12.2).[3] All the scientists shared a common characteristic: They were civilians, not military men. They were pacific persons, moderately idealists who liked nature or art, and that got involved, most by circumstances, in MP. All the men collaborated in the development of a bomb that, thanks to U/Pu chain reaction, presented maximum devastating power.

Albert Einstein
Old Grove Rd.
Nassau Point
Peconic, Long Island

August 2nd, 1939

F.D. Roosevelt,
President of the United States,
White House
Washington, D.C.

Sir:

Some recent work by E.Fermi and L. Szilard, which has been com-
municated to me in manuscript, leads me to expect that the element uranium may be turned into a new and important source of energy in the immediate future. Certain aspects of the situation which has arisen seem
to call for watchfulness and, if necessary, quick action on the part
of the Administration. I believe therefore that it is my duty to bring
to your attention the following facts and recommendations:

In the course of the last four months it has been made probable -
through the work of Joliot in France as well as Fermi and Szilard in
America - that it may become possible to set up a nuclear chain reaction
in a large mass of uranium,by which vast amounts of power and large quantities of new radium-like elements would be generated. Now it appears
almost certain that this could be achieved in the immediate future.

This new phenomenon would also lead to the construction of bombs,
and it is conceivable - though much less certain - that extremely powerful bombs of a new type may thus be constructed. A single bomb of this
type, carried by boat and exploded in a port, might very well destroy
the whole port together with some of the surrounding territory. However,
such bombs might very well prove to be too heavy for transportation by
air.

FIGURE 12.1   Continued

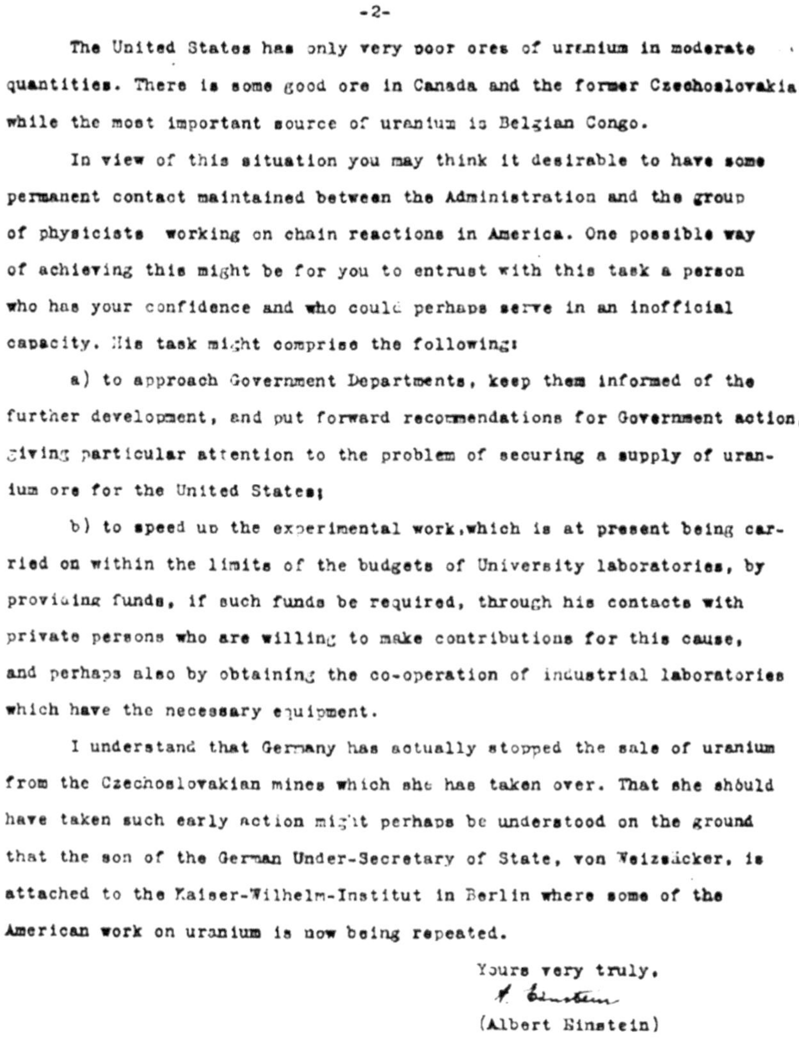

-2-

The United States has only very poor ores of uranium in moderate
quantities. There is some good ore in Canada and the former Czechoslovakia,
while the most important source of uranium is Belgian Congo.

In view of this situation you may think it desirable to have some
permanent contact maintained between the Administration and the group
of physicists working on chain reactions in America. One possible way
of achieving this might be for you to entrust with this task a person
who has your confidence and who could perhaps serve in an inofficial
capacity. His task might comprise the following:

a) to approach Government Departments, keep them informed of the
further development, and put forward recommendations for Government action,
giving particular attention to the problem of securing a supply of uranium ore for the United States;

b) to speed up the experimental work,which is at present being carried on within the limits of the budgets of University laboratories, by
providing funds, if such funds be required, through his contacts with
private persons who are willing to make contributions for this cause,
and perhaps also by obtaining the co-operation of industrial laboratories
which have the necessary equipment.

I understand that Germany has actually stopped the sale of uranium
from the Czechoslovakian mines which she has taken over. That she should
have taken such early action might perhaps be understood on the ground
that the son of the German Under-Secretary of State, von Weizsäcker, is
attached to the Kaiser-Wilhelm-Institut in Berlin where some of the
American work on uranium is now being repeated.

Yours very truly,

A. Einstein

(Albert Einstein)

FIGURE 12.1   Letter signed by Einstein to President Roosevelt.

The military head of MP had little to do with scientists at his charge.[3]
It was General Groves, a man of his time: He was openly racist, boastful
in his decision to use the nuclear bomb versus civil population, and
ignored President Truman in his declarations on explaining from who

was really the choice to drop the bombs. General Groves did not like to depend on *unreliable* civil scientists and maintained a high degree of control on all the persons that worked in MP, even revising private correspondence. Secrecy, censorship, and the total control, which prevailed in the installations, achieved that scientists extricated themselves from reality and went away from the exterior world (cf. Fig. 12.3). In spite of the police state created in MP installations, it is known certain persons of strong communist beliefs handed information abroad from MP, for example, Fuchs.

FIGURE 12.2 Top to bottom and left to right: Bethe, Oppenheimer, Teller, Fermi, Feynman, and Fuchs.

FIGURE 12.3 Exit poster of the installations of Oak Ridge of Manhattan Project by James E. Westcott.

After the first nuclear test (Trinity in honor of one of Oppenheimer's favorite poets, 1945), part of the scientific community of Los Alamos began to be frightened.[3] They were to be seen as creators of the weapon that could destroy humanity. Knowing that scientific knowledge will never be sole, they foresaw that the Soviets soon would have the bomb, and this could cause an arms race without precedents in history. The voices of the scientists were not listened and Oppenheimer was expelled from MP, because of his pacifist tendencies, and substituted by the other scientist with much more bellicose trends: Teller, who was of Hungary origin and naturalized in the United States. One of Teller's most criticized actions occurred during *Security Auditory*, promoted by Federal Bureau of Investigation, where he formulated a grave accusation versus Oppenheimer, denouncing him as a spy of communism. The charge was supported by Hoover, which caused Oppenheimer's exit of the scene, giving free way to Teller to coproduce H-bomb to which Oppenheimer was opposed. Teller, which was caricatured by Peter Sellers in Stanley Kubrick's film *Dr. Strangelove or: How I Learned to Stop Worrying and Love the Bomb* (1964) was, together with Ulam, the main inventor of H-bomb, in which U/Pu nuclear fission was used to start a reaction of H-nuclear fusion, much more energetic and destructive. The *H-bombs* nuclear test began in Pacific

(1952) with Ivy Mike, realizing this more than 10 million tons of 2,4,6-trini-trotoluene equivalent (Mton) of energy, the greatest bomb created till then in the first great nuclear fusion reaction carried out by humanity. Nuclear race followed its course before an expectant humanity and had its point of maximum terror and destructive-power demonstration in Soviet RDS-220 (*Tsar Bomba*) detonation (1964), in which bomb design Kurchatov and Sakharov participated, and 50 Mton were released in the greatest nuclear explosion till now. Without doubt, science, which till then brought inventions for humanity with the purpose of improving and enriching the life of persons, turned into a dark entity, which spoke a language unintelligible by most people and was the responsible for carrying fear and despair to the heart of humanity. The original sin with which science was marked will be forgotten with difficulty. Oppenheimer cited Indian poet Bhagavad-Gita gazing at Trinity nuclear mushroom cloud and referring to science: *Now I am become death, the destroyer of worlds.*

12.3   TRANSITION TO *ATOMS FOR PEACE* PROGRAM

Influenced by Oppenheimer's pacifist ideas, US President Eisenhower (1953), old general that fought in WW2 in Europe, made the speech AFP in the seat of the UN (cf. Fig. 12.4).[3] Trying that the United States did not go down in history of WW2 as a country that committed atrocities comparable to Nazi Germany, and being afraid that nuclear war break out in Europe, the exercise of reflection was carried out with which the pacific use of nuclear energy was promoted, not only in the United States but also in the whole world. The United States contributed equipment, information, and training to schools, hospitals, and research centers throughout the world on the civil use of nuclear energy. The first nuclear reactors of Iran, Israel, and Pakistan were built under the program of American Machine and Foundry.

FIGURE 12.4   *Atoms for Peace* symbol in 1955 *International Conference on the Peaceful Uses of Atomic Energy.*

A consequence of AFP, and the moral debate that developed around nuclear energy, was the establishment of International Atomic Energy Agency (IAEA, 1957), which seeks directly the protection of UN, and tries to promote the pacific use of nuclear energy and inhibit its military use (cf. Fig. 12.5).[3] In the attempt to endorse the peaceful utilization of nuclear energy, IAEA helps the Member States (171) to share scientific information, cooperate in technical aspects, and transfer knowledge on the civil use of nuclear energy. In the new stage, science seemed engaged in amending its acts via research and cooperation all over the world, and a model would be research in toroidal chambers with magnetic coils (*toroidal'naya kameras s magnitnymi katushkami, tokamaks*) and stella generators (*stellarators*).

FIGURE 12.5 US stamp of 1955 in allusion to *Atoms for Peace* Program.

12.4 *ATOMS FOR PEACE (1953)*

After Eisenhower (1953) AFP speech delivered to UN General Assembly and AEC Programme, Spanish traveling show (1958–, cf. Fig. 12.6) presented peaceful applications of nuclear energy to general public, with materials coming from AEC Programme. Roqué proposed paradox/hypothesis (P/H) on the conversion of the atom.[5]

P1. Nuclear science modernity contrasts with the reactionary ideology of Franco system.[6]

H1. There is no necessary historical relationship between science and democracy.

FIGURE 12.6 *Atoms for Peace* traveling show (1958–) presented nuclear energy's peaceful applications.

12.5 MAKING OF IGNORANCE VERSUS CONSCIENCE: OFFICIAL AND LOCAL SPEECHES

Howard informed American nuclear cover-up in Spain after Palomares (Almería, Spain) disaster (1966, cf. Fig. 12.7).[7] Florensa proposed hypotheses (H) on ignorance versus conscience-making in Palomares accident.[8]

H1. To reconstruct the processes of making of ignorance: the campaign of making of ignorance.

H2. (Fraga Iribarne). I did all that a father would do for their children.

FIGURE 12.7   Palomares (Almería, Spain) disaster (1966).

12.6   RADIOGRAPHY OF NUCLEAR WEAPONS

Peace Studies Delàs Centre organized round table Nuclear Weapons Radiography raising questions (Qs, cf. Fig. 12.8).[9]

Q1.   What threats do they arise in Trump era and the escalation of tensions in Northeast Asia?

Q2.   What are the possibilities of banning these weapons of mass destruction?

FIGURE 12.8   Peace Studies Delàs Centre organized round table nuclear weapons radiography.

NGO Peace Boat (PB, 1983–) proposed questions/answers (Q/A, cf. Fig. 12.9).

Q3. What is PB?

A3. PB is a Japan-based, international NGO that works since 1983 to promote peace/sustainability via organization of peace voyages onboard a large passenger ship.

Q4. How to pass victims' knowledge on to future generations/prevent recurrence of such tragedies?

Q5. To save people's blue planet with all living things on it as it is or go along self-destruction road?

Q6. Why does PB hesitate to prohibit nuclear weapons, which are far more destructive than biological/chemical ones?

A6. PB strong desire is to achieve a nuclear weapon-free world in victims' lifetime so that succeeding generations of people will not see hell on earth ever again.

FIGURE 12.9   Peace Boat's ship, the Ocean Dream, built in Denmark (1981), is the *floating peace village*.

12.7 CHERNOBYL, FUKUSHIMA, AND THE NUCLEAR POWER STATION OF COFRENTS

Montón and Hernàndez proposed questions/hypotheses/answer (Q/H/A) on Chernobyl, Fukushima, and Cofrents (València, Spain) nuclear power station of electric generation type boiling water reactor-6 (1984–, cf. Fig. 12.10).[10]

FIGURE 12.10   Cofrents nuclear power plant: general view.

Q1. Why did the accidents of Chernobyl (Ukraine, 1986) and Fukushima (Japan, 2011) occur?

Q2. What were the consequences for societies and environment?

Q3. What could it occur with an accident in the nuclear power station of Cofrents (cf. Fig. 12.11)?

FIGURE 12.11   Cofrents nuclear power plant cooling towers. Note not fume but water vapor over towers.

Q4. How do people live in Fukushima?

Q5. Individual temporary (?) store (ITS)?

H1. An unnecessary ITS.

Q6. Cofrents modified radioactive-waste spent fuel pool (2008), for greater storage capacity with ITS?

A6. Objective is to prolong its useful life till 60 years as they pretend in Garoña (Burgos)/Almaraz (Cáceres, Spain, cf. Fig. 12.12).

Q7. Infinite residues, why?

Q8. Does it make sense to generate in 40 years dangerous residues that will be active during hundreds thousands years?

Q9. In 200 years, will they really stay in the same conditions of safety?

Q10. In addition, in 2000 years?

H2. People do not need nuclear.

H3. Big energy companies realized that its business is outdated.

H4. Nuclear power stations are a barrier of entrance for renewable ones.

H5.   Cofrents = problems of safety.

Q11. Its nearness to Xúquer River downstream Alarcón (Cuenca)/
      Contreras (Cuenca/València, Spain) dams aggravates situation
      Vandellòs (Tarragona, Spain)/Fukushima/Chernobyl]; does it make
      sense?

H6.   When electricity price decays, users do not notice it; when it rises,
      they charge more although real cost be the same.

Q12. Will they want to prolong nuclear useful life to charge later for
      compensations if it is forced a closure before expiring end?

H7.   There are better alternatives.

H8.   An energy model is possible without risks, clean, economic, sustain-
      able, renewable, distributed, democratic and generating stable/
      quality employment.

FIGURE 12.12   Campaign *Close Down Cofrents*.

12.8   FINAL REMARKS

From the present results and discussion, the following final remarks can be drawn.

1. Nuclear science modernity contrasts with the reactionary ideology of Franco system. No necessary historical science–democracy relationship exists. American nuclear cover-up in Spain after Palomares disaster showed making of ignorance versus conscience via official versus local speeches. The objective is to reconstruct the processes of making of ignorance via its campaign.

2. Peace Boat is an international NGO that works to promote peace and sustainability via organization of peace voyages onboard a passenger ship. Activities must be based on the philosophy that any problem faced by any community is a global challenge that must be tackled via people–organizations–governments cooperation.

3. It is reasonable to assume that religious root, born of material conditions of every society, guided certain people by the way of placing growth avidity in front of harmony with environment, which caused fast development of important philosophical structures that rested on technical advance, but also placed the part of humanity that took the chief role in them in a limit situation, in which its own survival capacity is threatened. It is not waited that a dysregulated social evolution produce values that tend to solve the problem, before this turn irreversible, and an accorded policy, at international level and strictly measured, must be directed to promote the values and produce a pertinent legislation, which is the policy proposed in the report *Our Common Future*.

4. Concept *sustainable development* connects directly with West anthropocentric cultural tradition. An achievement is emphasizing, on rising them to the level of international political discussion, *synchronic and diachronic solidarity* concepts. Sustainable development needs to understand that Earth is a finite system, with a limited external power contribution, conditions that restrict productive model, and that economy is the tool that human being can handle to orientate development direction. Although different agents wanted to dispute the concept, this is clear in the terms in which it is defined in the report, being, however, actual policies still to be defined.

ACKNOWLEDGMENTS

The authors thank the support from Generalitat Valenciana (Project No. PROMETEO/2016/094) and Universidad Católica de Valencia *San Vicente Mártir* (Project Nos. UCV.PRO.17-18.AIV.03 and 2019-217-001).

KEYWORDS

- science
- Franco dictatorship
- technology
- nuclear energy
- Cold War
- Manhattan Project
- making ignorance
- nuclear weapon
- Chernobyl
- Fukushima
- Cofrents
- nuclear power plant

REFERENCES

1. Saraiva, T. *Fascist Pigs: Technoscientific Organisms and the History of Fascism*; MIT: Cambridge, MA, 2016.
2. Torrens, F.; Castellano, G. Nuclear Fusion and the American Nuclear Cover-Up in Spain: Palomares Disaster (1966). In *Engineering Technology and Industrial Chemistry with Applications*; Haghi, R., Torrens, F., Eds.; Apple Academic–CRC: Waretown, NJ, 2019; pp 297–308.
3. Fernández-Cosials, K. La Fusión Nuclear en Perspectiva (Chapter 6). In *Curso Básico de Fusión Nuclear*; Fernández-Cosials, K., Barbas Espa, A., Eds.; Jóvenes Nucleares–Sociedad Nuclear Española: Madrid, Spain, 2017.
4. Rhodes, R. *The Making of the Atomic Bomb: The 25th Anniversary of the Classic History*; Simon and Schuster: New York, 2012.
5. Roqué, X. La conversió de l'àtom: Ciències Nuclears i ideologia en el Franquisme. *Mètode* **2016,** *2016* (90), 77–83.

6. Soler Ferrán, P. *El Inicio de la Ciencia Nuclear en España*; Sociedad Nuclear Española: Madrid, 2017.

7. Howard, J. *White Sepulchres: Palomares Disaster Semicentennial Publication*; Universitat de València: València, Spain, 2016.

8. Florensa, C. In *Book of Abstracts, Science, Politics, Activism and Citizenship*, València, Spain, May 30–June 1, 2018; REDES CTS–Catalan Society for the History of Science and Technics: València, Spain, 2018; p O-21.

9. In *Book of Abstracts, Nuclear Weapons Radiography*, València, Spain, May 15, 2017; Peace Studies Delàs Centre: València, Spain, 2017.

10. Montón, R.; Hernàndez, F. J. *Txernòbil, Fukushima i la Central Nuclear de Cofrents*; Debats No. 31, Institució Alfons el Magnànim–CVEI–Diputació de València: València, Spain, 2017.

INDEX

T - #0843 - 101024 - C270 - 229/152/12 - PB - 9781774634684 - Gloss Lamination